Industrial Control Handbook
Volume 1:   Transducers

Industrial Control Handbook

Volume 1: Transducers

E.A. Parr
BSc, C.Eng., MIEE

COLLINS
8 Grafton Street, London W1

Collins Professional and Technical Books
William Collins Sons & Co. Ltd
8 Grafton Street, London W1X 3LA

First published in Great Britain by
Collins Professional and Technical Books 1986

Distributed in the United States of America
by Sheridan House, Inc.

British Library Cataloguing in Publication Data
Industrial control handbook.
　　Vol. 1 : Transducers
　　1. Process control
　　I. Parr, E.A.
　　670.42′7　　　　TS156.8

ISBN 0–00–383097–7

Typeset by David John Services Ltd, Maidenhead, Berks
Printed and bound in Great Britain by
Garden City Press, Herts

It is a capital mistake to theorise before one has the facts.

Sherlock Holmes in *Scandal in Bohemia*
Conan Doyle   1859–1930

Contents

Preface   *xiii*
Chapter 1.   Sensors and transducers   *1*
1.1. Instrumentation systems   1
1.2. Signals and standards   3
1.3. Definitions and terms   6
    1.3.1.   Introduction   6
    1.3.2.   Range and span   7
    1.3.3.   Linear and non-linear devices   7
    1.3.4.   Accuracy and error   9
    1.3.5.   Resolution   10
    1.3.6.   Repeatability and hysteresis   11
    1.3.7.   Environmental and ageing effects   12
    1.3.8.   Error band   13
1.4. Dynamic effects   14
    1.4.1.   Introduction   14
    1.4.2.   First-order systems   14
    1.4.3.   Second-order systems   17

Chapter 2.   Temperature sensors   *20*
2.1. Introduction   20
2.2. Principles of temperature measurement   21
2.3. Expansion thermometers   21
    2.3.1.   Coefficients of expansion   21
    2.3.2.   Bi-metallic thermometers   23
    2.3.3.   Gas-pressure thermometers   23
    2.3.4.   Vapour-pressure thermometers   25
2.4. Resistance thermometers   26
    2.4.1.   Basic theory   26
    2.4.2.   Resistance temperature detectors   27
    2.4.3.   Practical circuits   28

2.4.4. Thermistors	32
2.5. Thermocouples	36
2.5.1. The thermoelectric effect	36
2.5.2. Thermocouple laws	37
2.5.3. Thermocouple types	41
2.5.4. Extension and compensating cables	42
2.5.5. Cold junctions and cold junction compensation	45
2.5.6. Thermocouple temperature indicators	47
2.5.7. Using thermocouple tables	49
2.6. Radiation pyrometry	51
2.6.1. Introduction	51
2.6.2. Black-body radiation	52
2.6.3. Principles of pyrometry	55
2.6.4. Types of pyrometer	58
Chapter 3. Pressure transducers	62
3.1. Introduction	62
3.1.1. Definition of terms	62
3.1.2. Units	63
3.2. U-tube manometers	64
3.3. Elastic sensing elements	67
3.3.1. The Bourdon tube	67
3.3.2. Various sensing elements	68
3.3.3. Electrical displacement sensors	69
3.4. Piezo elements	73
3.5. Force balancing systems	76
3.5.1. Electrical systems	76
3.5.2. Pneumatic systems	77
3.6. Pressure transducer specifications	78
3.7. Installation notes	79
3.8. Vacuum measurement	82
3.8.1. Introduction	82
3.8.2. Pirani gauge	82
3.8.3. Ionisation gauge	83
Chapter 4. Position transducers	84
4.1. Introduction	84
4.2. The potentiometer	85
4.3. Synchros and resolvers	90
4.3.1. Basic theory	90
4.3.2. Synchro torque transmitter and receiver	91

4.3.3. Zeroing of synchros and lead transpositions   95
4.3.4. Control transformers and closed loop
        control   98
4.3.5. Differential synchros   101
4.3.6. Synchro identification   104
4.3.7. Resolvers   105
4.3.8. Solid-state converters   107
4.4. Shaft encoders   107
4.4.1. Absolute encoders   107
4.4.2. Incremental encoders   111
4.5. Small displacement transducers   116
4.5.1. Introduction   116
4.5.2. Linear variable differential transformer   116
4.5.3. Variable inductance transducers   119
4.5.4. Variable capacitance transducers   119
4.6. Proximity detectors   122
4.7. Integration of velocity and acceleration   124

Chapter 5.   Flow transducers   125
5.1. Introduction   125
5.2. Differential pressure flowmeters   126
5.2.1. Basic theory   126
5.2.2. Turbulent flow and Reynolds number   127
5.2.3. Incompressible fluids   128
5.2.4. Compressible fluids   130
5.2.5. Orifice plates   131
5.2.6. Dall tubes and venturi tubes   133
5.2.7. The Pitot tube   137
5.2.8. Measurement of differential pressure   140
5.3. Turbine flowmeters   142
5.4. Variable area flowmeter   144
5.5. Vortex shedding flowmeter   146
5.6. Electromagnetic flowmeter   148
5.7. Ultrasonic flowmeters   150
5.7.1. Doppler flowmeter   150
5.7.2. Cross-correlation flowmeter   153
5.8. Hot-wire anemometer   156
5.9. Injection flow measurement   157
5.10.Flow in open channels   158

Chapter 6.   Strain gauges, loadcells and weighing   159
6.1. First principles   159

6.1.1.	Introduction	159
6.1.2.	Stress and strain	161
6.1.3.	Shear strain	164
6.2.	Strain gauges	165
6.2.1.	Introduction	165
6.2.2.	Foil gauges	167
6.2.3.	Semiconductor strain gauges	170
6.3.	Bridge circuits	170
6.3.1.	Wheatstone bridge	170
6.3.2.	Temperature compensation	173
6.3.3.	Multigauge bridges	174
6.3.4.	Bridge balancing	176
6.3.5.	Bridge connections	176
6.3.6.	Bridge amplifiers	177
6.3.7.	Torque measurement	179
6.4.	Magnetoelastic devices	181
6.5.	Loadcells	183
6.6.	Weight controllers	187
6.7.	Belt weighers	189
Chapter 7.	*Level measurement*	*191*
7.1.	Introduction	191
7.2.	Float-based systems	193
7.3.	Pressure-operated transducers	197
7.3.1.	Direct measurement	197
7.3.2.	Gas reaction methods	200
7.3.3.	Collapsing resistive tube	201
7.4.	Direct electrical probes	202
7.4.1.	Capacitance sensing	202
7.4.2.	Resistive probes	204
7.5.	Ultrasonic methods	204
7.6.	Nucleonic methods	207
7.6.1.	Principles	207
7.6.2.	Radiation detectors	209
7.6.3.	Safety and legislation	211
7.7.	Level switches	213
Chapter 8.	*Optoelectronics*	*215*
8.1.	Introduction	215
8.2.	Electromagnetic radiation	215
8.3.	Optics	217
8.3.1.	Reflection and mirrors	217

8.3.2.	Refraction and lenses	221
8.3.3.	Prisms and chromatic aberration	225
8.4.	Sensors	227
8.4.1.	Photoresistors	227
8.4.2.	Photodiodes	229
8.4.3.	Phototransistors	230
8.4.4.	Photovoltaic cells	231
8.4.5.	Photomultipliers	231
8.4.6.	Integrated circuit devices	233
8.5.	Light emitters	233
8.5.1.	Light-emitting diodes	233
8.5.2.	Liquid crystal displays	235
8.5.3.	Incandescent emitters	237
8.5.4.	Atomic sources	238
8.6.	Fibre optics	239
8.7.	Photocells	243
8.8.	Lasers	247
8.9.	Miscellaneous topics	250
8.9.1.	Spectroscopy	250
8.9.2.	Flame failure devices	252
8.9.3.	Photometry	253
Chapter 9.	*Velocity, vibration and acceleration*	*256*
9.1.	Relationships	256
9.2.	Velocity measurement	257
9.2.1.	Tachogenerator	257
9.2.2.	Drag cup	260
9.2.3.	Pulse tachometers	261
9.2.4.	Tachometer mounting	264
9.2.5.	Doppler systems	265
9.3.	Accelerometers	267
9.3.1.	Seismic mass accelerometers	267
9.3.2.	Second-order systems	268
9.3.3.	Practical accelerometers	273
9.4.	Vibration transducers	277
Index		*279*

Preface

The topic of process control is usually covered from a very theoretical viewpoint. Although this mathematical background is important, it can lead to a false view of the role of the process control engineer in industry. The real-life engineer is employed by a firm that exists, or fails, on its ability to survive in the marketplace. Consequently the engineer is more concerned about acceptable (rather than perfect) control and factors such as reliability, ease of maintenance and 'whole-life' costs.

It is also worth noting that the process control engineer has to be the 'Compleat Engineer': master of electrical and electronic engineering, computing, hydraulics, chemistry, physics, pneumatics, plumbing and high finance as well as mathematician and analyst.

This series of books has been written as an introduction to process control for the undergraduate and newly qualified or practising engineer. It aims to be readable as well as instructive, and mathematics have been kept to a minimum (although the usual college theory is given in later sections).

The place at which most process control systems start is with some measurement of interest on an industrial plant. Volume 1 is therefore devoted mainly to the various quantities that need to be measured, and the commercial devices that are available for this purpose. Later volumes deal with specialised topics such as rotating machines, computing and the mathematical background to control engineering.

A book on transducers can never claim to be complete, and there will inevitably be some omissions. It is hoped that the devices described cover the vast majority of those found in industry. Notable omissions, no doubt, are transducers for chemical properties and wholly mechanical devices such as positive displacement flow meters. It may also be thought that description of light-emitting diodes and other light emitters does not fit in with the term 'transducers', but it was felt that 'optoelectronics' should be covered as one topic.

The plague of the process control industry is undoubtedly a lack of consistency in the choice of units. The author has yet to see a wholly consistent plant, and most have weird meters calibrated in, say, kilograms per square foot. SI units have been used throughout this series for the most part, but the reader will find the odd lapse into p.s.i., gallons or feet where this seems more in line with industrial practice. The reader is reminded of Emerson's quote: 'A foolish consistency is the hobgoblin of little minds.'

There are many manufacturers who have assisted with this book by providing data sheets, application notes, helpful suggestions and photographs. These are acknowledged in the relevant places in the book. Special thanks are also due to my own employers, the Sheerness Steel Company, for facilities and for their interest in the project.

Extra special thanks must go to my wife Alison for doing all the typing while managing to run the house (and a part-time job) and tolerating the mountains of reference material in most rooms. The rest of my family (Simon, Jamie and Nick) are also due thanks, and apologies, for many weekends and evenings when 'Daddy's working'.

Andrew Parr
Isle of Sheppey,
Kent

September 1985

Chapter 1
Sensors and transducers

1.1. Instrumentation systems

Accurate measurement of strategic quantities such as flow, pressure or temperature is an essential part of the control or monitoring of the operation of any process. Figure 1.1 represents most industrial instrumentation systems. A physical quantity, called the *measurand* or *process variable* is converted by a measurement system into a measured value, usually an electrical or pneumatic signal, which can be used for display or control. The first volume of this series is concerned with the conversion from process variable to measured value.

Figure 1.1 can be redrawn in more detail as fig. 1.2. A sensing element, or *sensor,* is connected to the process and experiences a change which relates to the process variable being monitored. A platinum resistance thermometer, for example, experiences a change in resistance with temperature, or a flow-dependent differential pressure is developed across an orifice plate.

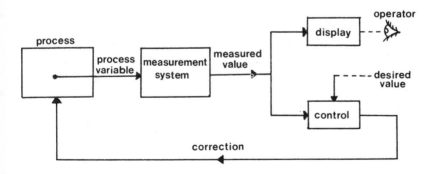

Fig. 1.1   Industrial instrumentation and control system.

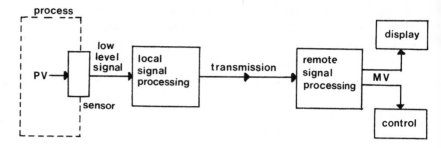

Fig. 1.2 Elements of an instrumentation system.

In many systems there may be a chain of sensing elements to measure just one variable. An orifice plate is used to measure flow, and produces a flow-dependent differential pressure. This pressure may be converted to a positional displacement by bellows, and then the displacement is converted to an electrical signal by a potentiometer. In such systems the first sensor is called the *primary sensor* (the orifice plate in our example).

The initial signal directly off the sensor is often small, so local signal processing or conditioning may be used. Typical signal conditioning is a bridge circuit to convert the change in resistance of a strain gauge to an electrical voltage, or a frequency-to-voltage converter to change the flow-dependent pulse train from a turbine flow meter to a flow-dependent voltage. Other simple processing examples are filtering and linear amplification. More complex processing could take place where a radio telemetry transmission path is used. The requirement here would probably be multiplexing analog to digital conversion and some form of modulation.

The transmission path from process to display/control device is of prime importance, as it will inevitably introduce errors into the measurement system. Examples of transmission paths are electrical cables (for voltage and current), pneumatic pipes, fibre optic cables and radio links. Errors are introduced via interference (noise) and cable impedances into electrical systems, and by minute, but unavoidable, leaks in pneumatic systems. All transmission paths also inherently introduce a lag into the system; the measured value cannot react instantaneously to changes in the process variable. Capacitance and inductance cause lags in electrical systems, and finite pipe volumes lags in pneumatic systems.

In most systems further signal processing is performed local to the display or control device. In telemetry, obviously, a receiver and demodulator are required. Often linearisation is performed at this point (e.g. square-root conversion for differential pressure flow meters, described later in chapter 5). Some form of computational correction may also be performed; applying cold junction compensation for thermocouples or temperature correction for mass flow measurements are typical examples.

The term *transducer* is often encountered. In strict terms a transducer is a device that converts one physical quantity into another, the second being an analog representation of the first. A thermocouple is a transducer that converts temperature to an electrical potential. It is more common, however, to use the term sensor for the actual measurement device (i.e. the primary sensor) and transducer for the entire measuring system local to the plant (including local signal processing). There are, however, no strict rules, and in many cases the terms sensor and transducer are used interchangeably. The word *transmitter* is also often used to mean transducer or sensor.

1.2.   Signals and standards

Primary sensors produce a wide variety of signals: strain gauges give a very small resistance change, resistance thermometers a larger resistance change, thermocouples a voltage of a few millivolts, position measuring potentiometers several volts, and so on. Commercial transducers (with the word 'transducer' used to indicate signal conditioning as described in the previous section) are designed to give standard output signals for transmission to the control and display devices.

There are obvious maintenance and design advantages to standardised signals. If all the instrumentation, say, is designed around one standard, there can be commonality of spares and no need for specialised fault-finding aids.

The commonest electrical standard is the 4 to 20 mA current loop. As its name implies, this uses a variable current signal with 4 mA representing one end of the signal range and 20 mA the other. The current loop is totally floating from earth (and will not work correctly if an earth is applied to the signal lines). This gives excellent noise immunity as common mode noise has no effect and errors caused by different earth potentials around the plant are avoided. Because current, rather than voltage, is used, line resistance has no effect.

Several display/control devices can be connected *in series* (as in fig. 1.3b), providing the total resistance does not rise above some value specified for the transducer (usually around 1 kohm).

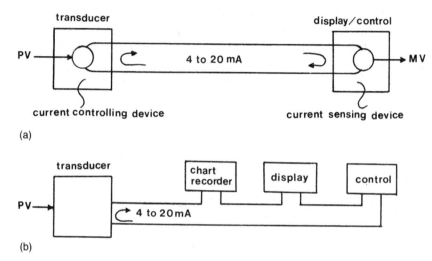

Fig. 1.3   The 4 to 20 mA current loop transmission system. (a) Principle of 4 to 20 mA current loop. (b) Series connection of display/control devices.

Transducers using 4–20 mA can be current sourcing (with their own power supply, as in fig. 1.4a) or designed for two-wire operation where the signal wires also act as the power supply connections, as in fig. 1.4b. In the latter (and obviously cheaper to install) case, a separate power supply local to the display provides, say, 24 V DC. The transducer senses the current being drawn, and adjusts a shunt regulator to give the correct current for the process variable while maintaining sufficient voltage for its own internal electronics. The loop current then returns through the display device to the negative side of the power supply. Many commercial controllers incorporate a suitable power supply to allow self-powered or two-wire transducers to be used. Alternatively, one power supply can feed several transducers with some loss of isolation. Voltage-to-current and current-to-voltage circuits are often based on the DC amplifier circuits of fig. 1.4c, d. These circuits are discussed further in chapter 1, Volume 2.

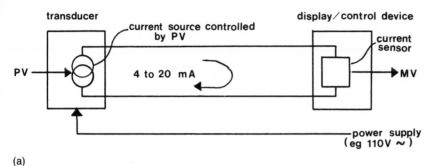

(a)

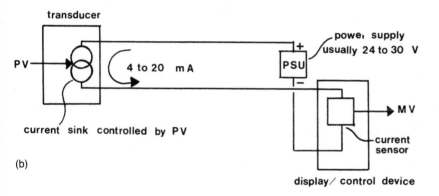

(b)

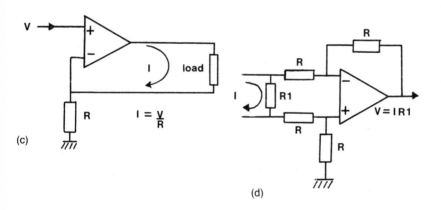

(c)

(d)

Fig. 1.4 Current loop circuits. (a) Self-powered transducer. (b) Two-wire operation. (c) Voltage-to-current conversion. (d) Current-to-voltage conversion.

The use of an offset zero (4 mA) has several advantages (not least of which is the provision of sufficient current for a two-wire transducer to continue working at 'zero' output). If a zero voltage or current for the bottom of the range was chosen, an open-circuit or short-circuit line would look like a bottom-range signal. Any line fault on a 4–20 mA line will cause a substantial 'negative' signal which is easily detected at the controller or display device. In addition, the signal is decidedly unipolar, giving no ambiguities around zero, and obviating the need for a negative power supply which would be needed to give a zero voltage or current output.

Although 4–20 mA is by far the commonest electrical standard, others may be encountered. Among these are 10–50 mA and 1–5 V, again using an offset zero. Often 4–20 mA signals are converted to 1–5 V at the display device or controller by a series 250 ohm resistor.

Pneumatic signals also use an offset zero, the commonest standards being 3–15 p.s.i. or its metric equivalent, 0.2–1 bar (20–100 kPa). An offset zero improves the speed of response of a pneumatic system as well as bringing similar advantages to an electrical offset zero. A pneumatic line loses pressure by venting to atmosphere, and follows an exponential decay curve with respect to time. The use of an offset zero reduces the time taken to go from full scale to zero.

1.3. Definitions and terms

1.3.1 Introduction

The process variable, PV, of fig. 1.5 is measured by some instrumentation system to give a measured value, MV, which is an analog representation of the value of PV. There will, inevitably, be errors in this representation. Perfect representation can never be achieved, but there will always be a maximum allowable error. A temperature measurement in a chemical process, for example, may need to be accurate to $\pm 2°C$, or the level measurement in a vat can tolerate a ± 10 mm error. The process control engineer needs to be able to predict the errors in a measurement system in order to ensure that adequate accuracy is obtained but that money is not wasted on a system that is unnecessarily accurate.

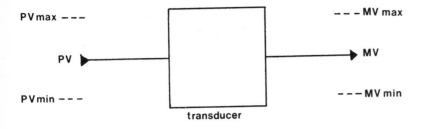

Fig. 1.5   Range and span of a transducer.

Transducer specifications use every-day terms like accuracy, error and repeatability in a precise way. The rest of this section defines the terms used on transducer data sheets.

1.3.2.   Range and span

The *range* of a value is specified by its maximum and minimum values. On fig. 1.5 the input range is PV_{min} to PV_{max}, and the output range MV_{min} to MV_{max}. A pressure transducer, for example, could have an input range of 0–100 kPa, and an output range of 4–20 mA. A single sensor can be similarly defined; a type K thermocouple could have an input range of 200–500 °C and an output range of 8–20 mV.

The *span* is the difference between maximum and minimum values: the input span is $(PV_{max} - PV_{min})$ and the output span $(MV_{max} - MV_{min})$. For example, the type K thermocouple just mentioned has an input span of 300 °C and an output span of 12 mV.

1.3.3.   Linear and non-linear devices

If the relationship between MV and PV is plotted on a graph, a result similar to fig. 1.6a will probably be obtained. The ideal relationship will be a straight line, fig. 1.6b, which has the form:

$$MV = K \cdot PV + Z \tag{1.1}$$

where K is the *sensitivity* or *scale* factor given by:

$$K = \frac{MV_{max} - MV_{min}}{PV_{max} - PV_{min}} \tag{1.2}$$

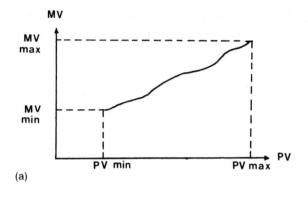

(a)

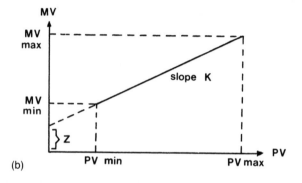

(b)

Fig. 1.6   Relationship between MV and PV. (a) Typical transducer response. (b) Idealised transducer response.

K will have the units of MV/PV (e.g. mA/kPa for a pressure transducer with current output). Z is the zero offset given by:

$$Z = MV_{min} - K \cdot PV_{min} \tag{1.3}$$

Z can be positive or negative.

A device that can be represented with tolerable error by equation 1.1 is said to be a linear device. If the relationship between PV and MV cannot be represented by equation 1.1, it is said to be non-linear. Devices with known non-linearities can be made linear by suitable signal conditioning. An orifice plate, for example, has a square law response between flow (PV) and differential pressure (MV), i.e.:

$$MV = A \cdot PV^2 \tag{1.4}$$

where A is a constant. If the orifice plate signal is passed through a circuit that performs a square-root operation, a linear relationship is obtained. Many devices can be linearised by considering them to be of the form:

$$MV = A + B\ ^*PV + C^*PV^2 + \ldots \qquad (1.5)$$

where A, B, C, etc. are constants (which can be positive or negative), and designing a suitable compensating circuit.

1.3.4. Accuracy and error

The *accuracy* of an instrument is a measure of how close the measured value is to the process variable. Accuracy is a rather loose term, and the precise term *error* is more generally used. This is defined as the maximum difference which may occur between the process variable and the measured value.

Error can be expressed in many ways. The commonest are absolute value (e.g. the maximum error on a temperature measurement may be defined as $\pm2°C$ regardless of value), as a percentage of actual value of the process variable, or as a percentage of full scale of the measuring device (usually called FSD for full-scale deflection). The device in fig. 1.7, for example, is a pressure-measuring transducer with an exaggerated non-linearity. It has an absolute maximum error of 0.5 p.s.i.,

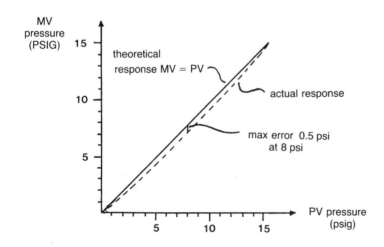

Fig. 1.7 Transducer error as difference between ideal and actual response. Note that MV is expressed in the same units and range as PV, although in practice MV would be an electrical signal of, say, 4 to 20 mA.

which can be expressed as 6.25% of actual value or 3.33% of FSD.

1.3.5. Resolution

Many devices have an inherent 'coarseness' in their measuring capabilities. A wirewound potentiometer, used for position measurement as in fig. 1.8a, has an inherent step size determined by the gauge of wire used. This gives a response similar to fig. 1.8b. Even a cermet type pot has an ultimate coarseness caused by the grain size of the material used.

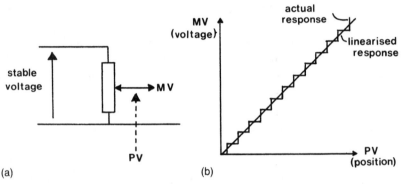

Fig. 1.8 Transducer with finite resolution. (a) Potentiometer circuit: PV is movement of slider, MV is output voltage. In theory PV = K × MV where K is a constant. (b) Actual response showing finite resolution.

Totally digital systems (e.g. shaft encoders or equipment using analog-to-digital and digital-to-analog converters, see chapter 3 in Volume 2) also have an easily definable step-like response. Less obviously, almost all supposedly analog systems come up against some limit which prevents a more precise reading being made.

The term *resolution* is used to define the 'steps' in which a reading can sensibly be made. The resolution can be larger or smaller than the inherent error. A well-constructed steel ruler kept in a temperature-controlled environment may have an absolute error of less than ±0.1 mm. A cheap plastic ruler may have an absolute error of more than ±2 mm. If read by eye, the resolution of both would, however, be the same at about ±0.5 mm, or half a scale division. Resolution and error are additive in terms of overall *system* error. The measuring accuracy of the plastic ruler is ±2.5 mm.

Resolution becomes important when comparisons are to be made. A low-accuracy device with good resolution can be used to compare two values or to indicate if a value is increasing or decreasing. The cheap plastic ruler above, for example, could indicate the longer of two items with a resolution of 0.5 mm.

1.3.6 Repeatability and hysteresis

In many applications, the accuracy of a measurement is of less importance than its consistency. Where material is cut to some length, for example, the consistency of cutting accuracy may be of more importance than the absolute accuracy. Similarly a process may be required to keep at the *same* temperature for some period, but the actual temperature need not be known to any great degree of accuracy. The consistency of a measurement is defined by the terms repeatability and hysteresis.

Repeatability is defined as the difference in readings obtained when the same measuring point is approached several times from the *same* direction.

Hysteresis occurs when the measured value depends on whether the process variable approached its current value by increasing or decreasing its previous value. The commonest example, backlash in gears or mechanical linkages, is illustrated in fig. 1.9, giving a hysteresis error, h, between increasing and decreasing readings. Stiction (where a certain minimum force is needed to move an object) is also a common cause of hysteresis.

The effect of hysteresis can be reduced, or eliminated, by careful system design. The use of sprung gears (effectively two gear wheels on the same shaft tensioned by a spring) and similar pre-tensioning of couplings can remove mechanical hysteresis. An alternative approach, common in position control systems, is to use a unidirectional approach as in fig. 1.9c. At time T the position is required to go from C to A which is a reversal, so the system introduces a deliberate overshoot to allow position A to be approached in the same direction as the other previous movements.

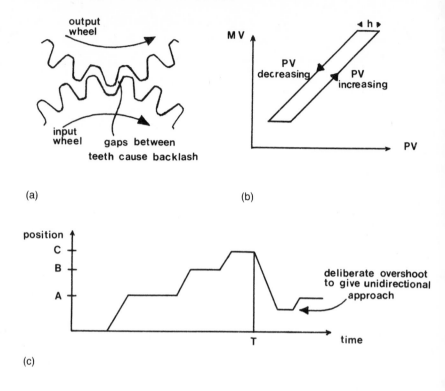

Fig. 1.9 Causes and effects of hysteresis. (a) Gearwheels with exaggerated play. (b) Effect of backlash. (c) Eliminating the effect of hysteresis in a position control system by means of unidirectional approach.

1.3.7. Environmental and ageing effects

The specifications for a transducer normally detail the possible errors outlined above for a new calibrated instrument under fixed operating conditions (e.g. constant ambient temperature, fixed electrical supply voltage or instrument air supply pressure). The accuracy of the transducer will be adversely affected by changes in its environment, and will progressively degrade with age. Both these effects will manifest themselves as a zero shift (or zero error), as in fig. 1.10a, or a sensitivity change (or span error), as in fig. 1.10b. Both may change with time, an effect known as *drift*.

Environmental effects are usually defined as the percentage error for some environmental change. A differential pressure transmitter, for example, may be affected by the static pressure.

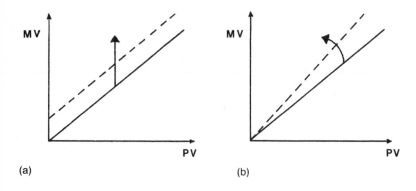

Fig. 1.10 The effects of environmental changes and ageing. (a) Zero shift. (b) Span shift.

A transducer with a 4–20 mA current output may be slightly load dependent. Almost all instruments are ambient temperature sensitive to some extent. Specifications list the possible error sources for a transducer, allowing the process control engineer to assess their importance in a particular application. Temperature effects may be crucially important for plant-mounted equipment, which may experience ambient changes from −15 °C to +50 °C, but are of little relevance for equipment in an air-conditioned control room.

Known environmental effects can be eliminated by the inclusion of suitable compensation. The commonest example of this is probably the use of cold junction compensation with thermocouples: see section 2.5.5.

Age effects are less important with modern solid-state electronics-based instruments than with earlier vacuum-tube or pneumatic devices. Age effects can, however, be eliminated by planned maintenance and recalibration at regular intervals. Device specifications will, again, allow the engineer to devise suitable schedules to keep age effects within acceptable limits.

1.3.8. Error band

In many instruments, all the effects of sections 1.3.3–1.3.7 will be individually small and possibly difficult to measure. It is therefore becoming increasingly common for manufacturers to specify a *total error*, or *error band*, figure which includes *all* the above effects. This can be specified in the same way as error was defined earlier; as absolute or percentage of FSD (fig. 1.11a) or as a percentage of value (fig. 1.11b). In all cases the measured value lies within the specified limits.

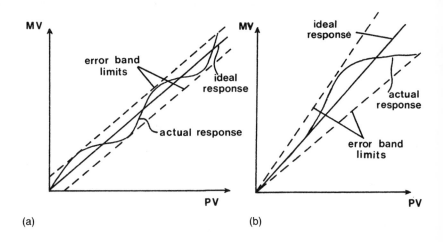

Fig. 1.11 Definitions of total error and error band specifications. (a) Absolute value or percent FSD. (b) Percentage of value.

1.4. Dynamic effects

1.4.1. Introduction

The definitions in section 1.3 refer to the static characteristics of transducers, i.e. the measurements of signals which are considered not to vary. In many applications the process variable will be changing rapidly and it will be necessary for the transducer to follow these variations. A pressure transducer, for example, may be needed to follow the pressure variations inside the cylinders of an internal combustion engine where pressure spikes of less than 1 ms must be recorded.

1.4.2. First-order systems

If a temperature sensor at an ambient temperature of 20 °C is suddenly plunged into water at 80 °C, it will experience a step change of process variable as shown in fig. 1.12a. The indicated temperature (ignoring errors) will, however, follow a curve as shown in fig. 1.12b, the delay being caused by the time taken for the sensor to heat up to the temperature of the water.

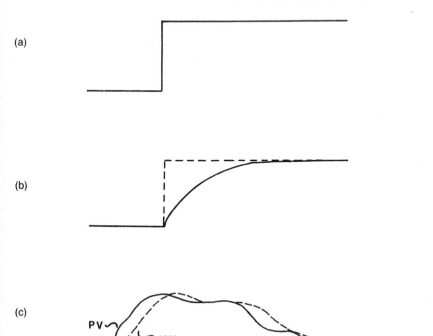

(a)

(b)

(c)

Fig. 1.12 Response of first-order system. (a) Step change in PV. (b) Observed change in MV. (c) Response to slow-changing PV.

Curves similar to fig. 1.12b arise when the rate of change of the output is proportional to the difference between the current value, V, and the final value, V_f. This obviously applies to many thermal transducers. We can write:

$$\frac{\mathrm{d}V}{\mathrm{d}t} = K(V_f - V_i) \tag{1.6}$$

where K is a constant.

Solving equation 1.6 gives:

$$V = V_f - (V_f - V_i) \exp(-Kt) \tag{1.7}$$

This can be more conventionally expressed for practical purposes by replacing the constant K by $1/T$ where T is called the *time constant*. Equation 1.7 then becomes:

$$V = V_f - (V_f - V_i)\exp(-t/T) \tag{1.8}$$

The term $\exp(-t/T)$ approaches zero as t becomes large, so V approaches V_f as shown in fig. 1.12b. Equation 1.8 is the response of a *first-order linear system*.

The time constant T determines the delay, V reaching 63% of V_f in a time T. The table that follows shows the percentage difference reduced for various multiples of T after a step change of input.

Time	Percentage difference reduced
T	63
$2T$	86
$3T$	95
$4T$	98
$5T$	99

It follows that for a step change in input, significant errors in measured value will occur. If the temperature transducer in fig. 1.12 has a time constant of 4 s, the indicated temperature will be:

Time(s)	Reading (°C) (percentage from table above *60 + 20)
4	57.8
8	71.6
12	77
16	78.8
20	79.4

A step change is a worst-case input, but even a more natural input (fig. 1.12c) will give a significant delay. It is good practice, however, to design a system such that a step-change signal gives an acceptable delay in the knowledge that all real inputs will comfortably meet the design criteria.

1.4.3. Second-order systems

A second-order response occurs when the transducer contains an element which is analogous to a mechanical spring/viscous damper or an electrical tuned circuit. The response of such a system to a step input of height a is given by the second-order equation:

$$\frac{d^2x}{dt^2} + 2b\omega_n\frac{dx}{dt} + \omega_n^2x = a \tag{1.9}$$

where b is called the *damping factor* and ω_n is called the *natural frequency*. The final value of x is given by:

$$x = a/\omega_n^2 \tag{1.10}$$

The step response depends on both b and ω_n, the former determining the overshoot and the latter the speed of response as shown on fig. 1.13a. It can be seen that for values of $b<1$ damped oscillations occur causing a first overshoot as in fig. 1.13b. The time to the first overshoot is given by:

$$T = \frac{1}{2f_n\sqrt{(1 - b^2)}} \tag{1.11}$$

where $f_n = \omega_n/2\pi$. For $b>1$ no overshoot occurs. The special case $b = 1$ is called critical damping and represents the fastest response without overshoot for a given ω_n.

If an overshoot can be tolerated, other values of b may be advantageous. Figure 1.14 shows the Bode diagrams for a second-order system (normalised for ω/ω_n). In most applications, the fastest possible response will be required from a transducer, and this implies as wide a frequency response as possible.

An ideal response will remain within a given deviation from a normalised unity gain to as high a frequency as possible. Examination of fig. 1.14a shows that this occurs for $b = 0.7$ (and not $b = 1$ as might first be thought). Similarly, fig. 1.14b shows that the phase shift for $b = 0.7$ is lower than for $b = 1$ for frequencies below ω_n (and the relationship between ϕ and ω is more linear). Figure 1.13b shows that $b = 0.7$ gives an overshoot of less than 10%. If this can be tolerated, a selection of $b = 0.7$ gives the best response, and many transducers are specified with this value of damping.

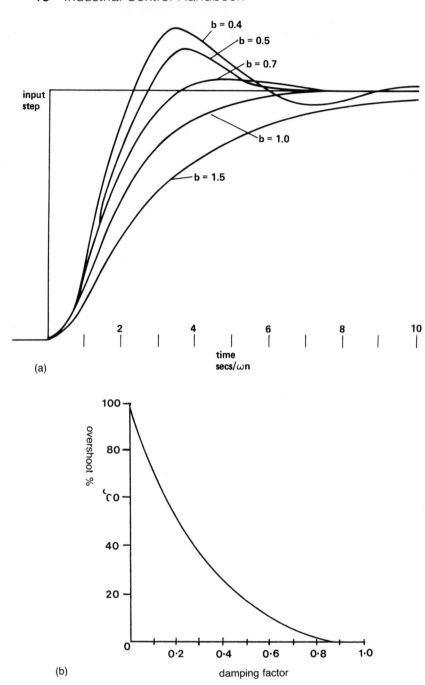

Fig. 1.13 Effect of different damping factors. (a) Step response for various damping factors. (b) Overshoot related to damping factor.

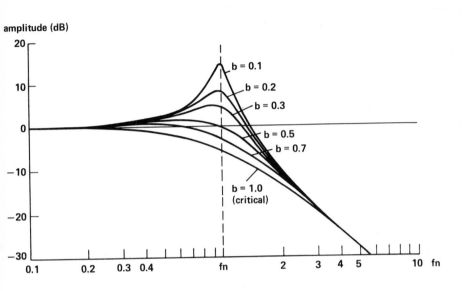

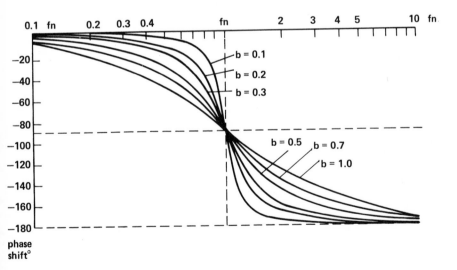

Fig. 1.14   Bode diagrams for second-order system.

Chapter 2
Temperature sensors

2.1. Introduction

The measurement and control of temperature are possibly the most common operations in process control. To measure temperature qualitatively we need to define a temperature scale. This is done by choosing two temperatures at which some readily identifiable physical effect occurs, and assigning numerical values to these temperatures. Other temperatures can then be found by interpolation.

The Fahrenheit and Celsius (or Centigrade) scales use the freezing and boiling points of water as the two reference points.

	Fahrenheit	Celsius
Freezing point	32	0
Boiling point	212	100

It follows that $F = (9.C)/5 + 32$ and $C = (F - 32).9/5$.

The SI unit of temperature is the Kelvin (it should be noted that, unlike °F and °C, the degree symbol (°) is not used on the Kelvin scale). The lower of the two defining points on the Kelvin scale is absolute zero. This is the lowest theoretical temperature and is defined as 0 K. For comparison, $0 \text{ K} = -273.16 \,°C$. The second defining point is the triple point of water (this being the unique temperature at which gaseous, liquid and solid phases can be in equilibrium). This is defined as 273.16 K, which corresponds to 0.01 °C. The definition is chosen such that a change in termperature of 1 K corresponds to a change of 1 °C and

$$K = °C + 273.15$$

In industrial applications, the Celsius scale is most widely used, but conversion to the Kelvin scale is often needed, for example when gaseous volumes or pressures are converted to some standard pressure.

2.2. Principles of temperature measurement

There are, in general, four types of temperature sensor based on the following physical properties, which are temperature dependent:

(1) Expansion of a substance with temperature, which produces a change in length, volume or pressure. In its simplest form this is the common mercury-in-glass or alcohol-in-glass thermometer.
(2) Changes in electrical resistance with temperature, used in resistance thermometers and thermistors.
(3) Changes in contact potential between dissimilar metals with temperature; thermocouples.
(4) Changes in radiated energy with temperature; optical and radiation pyrometers.

Transducers based on these principles will be described in following sections.

2.3. Expansion thermometers

2.3.1. Coefficients of expansion

In fig. 2.1a we have a rod of length L_0 at some temperature T_0. If this is heated to some higher temperature, T_1, the rod will increase to a new length L_1 given by:

$$L_1 = L_0(1 + \gamma(T_1 - T_0)) \tag{2.1}$$

where γ is defined as the coefficient of linear thermal expansion. (The relationship does, in reality, include terms in $(T_1 - T_0)^2$ and higher terms, but the above equation is accurate for all practical purposes.)

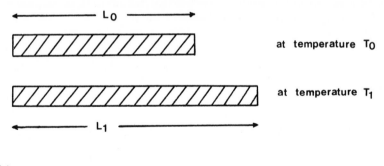

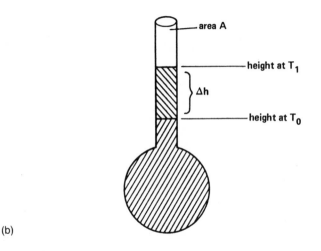

(a)

(b)

Fig. 2.1 Expansion of liquids and solids with temperature. (a) Expansion of a solid rod with temperature. (b) Expansion of liquid with temperature.

Typical values of γ are, per degree Celsius:

Steel	6.7×10^{-6}
Copper	16.6×10^{-6}
Aluminium	25×10^{-6}

In fig. 2.1b, a volume V_0 of liquid is at temperature T_0. If this is heated to temperature T_1, its volume will increase to V_1, again given by:

$$V_1 = V_0(1 + \alpha(T_1 - T_0)) \tag{2.2}$$

where α is the coefficient of cubical thermal expansion. Again, the full equation has higher-order terms which may be neglected. Typical values of α are, per degree Celsius:

Mercury	0.56×10^{-4}
Alcohol	0.35×10^{-3}

In fig. 2.1b, the increase in volume ΔV appears as a change Δh in the length of the liquid column in a capillary tube. If A is the cross-sectional area of the tube:

$$\Delta h = \frac{\Delta V}{A} \qquad (2.3)$$

If A is made small, significant changes in the column length can be obtained for small changes in temperature. This is the basis of mercury-in-glass and alcohol-in-glass thermometers. In practice, a small correction needs to be made for the expansion of the container holding the liquid.

2.3.2. Bimetallic thermometers

In fig. 2.2a, two dissimilar metals, A and B, have been bonded together. Metal A has a high coefficient of expansion and metal B a low coefficient of expansion. (The alloy Invar is often used for metal B.) As the temperature rises, the greater expansion of metal A causes the bar to bend, producing a deflection d which is a function of temperature. The change, d, is small in fig. 2.2a, but can be increased and made more linear by the use of a coiled bimetallic spring as shown in fig. 2.2b.

Bimetallic thermometers are cheap but of quite low accuracy. They are not widely used in industry, mainly because they cannot provide remote indication. Temperature sensing switches (thermostats) are often based on fig. 2.2b with switch contacts replacing the pointer.

2.3.3. Gas-pressure thermometers

If a gas is contained in a vessel, Charles's law states that:

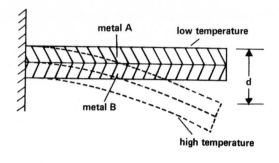

(a)

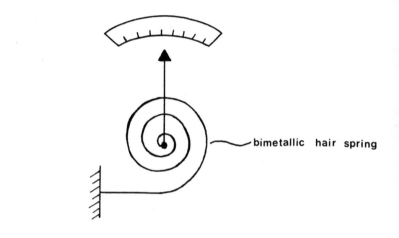

(b)

Fig. 2.2 Bimetallic thermometers. (a) Bimetallic strip. (b) Indicating bimetallic thermometer.

$$\frac{PV}{T} = \text{constant} \tag{2.4}$$

where P is the gas pressure, V is the gas volume, and T is the gas absolute temperature. If V is constant (i.e. a sealed container), then

$$P = \alpha T \tag{2.5}$$

where α is a constant, dependent on the gas and the initial pressure in the container. The equation implies that the pressure is linear with absolute temperature.

Gas-pressure thermometers can be used with remote indication provided the volume of the capillary tube connecting the bulb to the indicator is small by comparison with the bulb itself. A typical instrument, using nitrogen, can achieve an accuracy of 2% while giving remote indication up to 25 m from the bulb. Because no electrical power is used, they are particularly suited for use in hazardous areas.

2.3.4. Vapour-pressure thermometers

Some of the restrictions of the gas-pressure thermometer can be overcome by using the vapour pressure of a volatile liquid. Methyl chloride, for example, has the vapour-pressure curve of fig. 2.3a. As can be seen, there is a 1 : 10 pressure variation over a 100°C temperature variation.

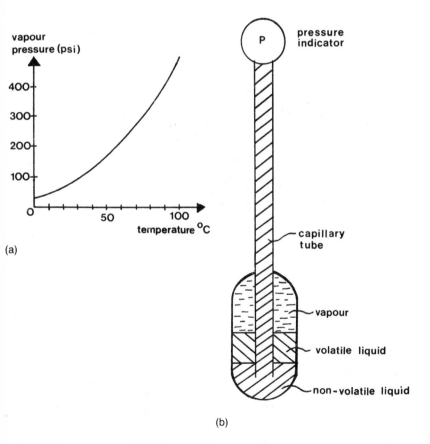

(a)

(b)

Fig. 2.3 Vapour-pressure thermometers. (a) Vapour-pressure curve for methyl chloride. (b) Practical thermometer.

The construction of a typical instrument is shown in fig. 2.3b. The volatile liquid is entrapped in the bulb via a non-volatile liquid which also fills the capillary tube to the indicator. The vapour pressure is thus relayed directly to the indicator, and the temperature or volume of the capillary tube has no effect.

The major constraint on the distance between bulb and indicator is speed of response. A practical limit is about 100 m, at which the instrument will exhibit a time constant of about 20 s. Like the gas-pressure instrument, the vapour-pressure thermometer is well suited to hazardous environments.

2.4. Resistance thermometers

2.4.1. Basic theory

The electrical resistance of most metals increases approximately linearly with temperature. If a metal wire has a temperature R_0 at 0 °C, then the resistance at T °C will be given by:

$$R_T = R_0 (1 + \alpha T + \beta T^2 + \ldots) \qquad (2.6)$$

In almost all industrial applications, terms higher than the square can be ignored, and for most the relationship

$$R_T = R_0 (1 + \alpha T) \qquad (2.7)$$

will suffice if $0 < T < 150$ °C.

The constant α is called the temperature coefficient of resistance. Typical values are:

Metal	α
Platinum	0.0039
Copper	0.0043
Nickel	0.0068

Figure 2.4 shows the variations in resistance with temperature for various metals. In each case, a resistance of 100 ohm at 0 °C is used as a reference. As can be seen, a nickel-based thermometer is most sensitive, but is the most non-linear (due to the βT^2 and higher terms). Platinum is the least sensitive, but the most linear. In practice, the choice of material for a specific application will be determined by the accuracy required, the ability to resist

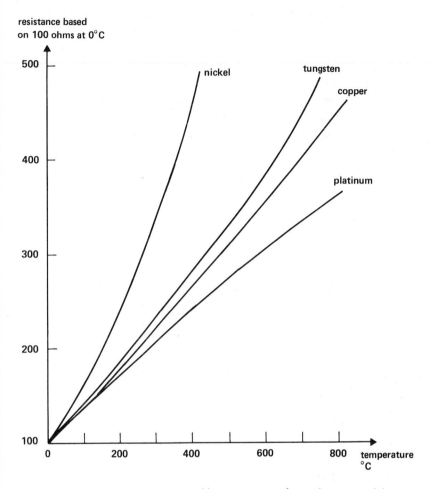

resistance based
on 100 ohms at 0°C

Fig. 2.4 Change of resistance with temperature for various materials.

contamination, and the cost. Platinum-based resistance thermometers are probably the most widely used (although they are the most expensive).

2.4.2. Resistance temperature detectors

A temperature transducer based on the above principles is called a resistance temperature detector (RTD), and is specified in terms of its resistance at 0 °C, and the change in resistance from 0 °C to 100 °C. This is known as the *fundamental interval*.

Platinum RTDs are constructed with a resistance of 100 ohm at 0 °C (and are often referred to as PT100 sensors). This gives a resistance of 138.5 ohm at 100 °C, and hence a fundamental interval of 38.5 ohm. In the UK, the relevant standard for RTDs is BS 1904 which specifies calibration methods and tolerances for the sensor. PT100 sensors can be used over a temperature range of −200 °C to 800 °C with an accuracy of ±0.5% between 0 °C and 100 °C and ±3% at the extremes of the temperature range.

RTDs are available in many shapes and sizes; a typical sensor is shown in fig. 2.5. The construction is a trade-off between protection against the atmosphere or fluid whose temperature is to be measured, and physical size (which determines the time constant of the RTD's response to a temperature change).

Fig. 2.5 Examples of RTD sensors.

Figure 2.6 shows various constructions of RTDs. These are designed to protect the wire from mechanical shock while not applying any stress on the wire (which would cause resistance changes in a similar way to a strain gauge). Constructions with the wire in direct contact with the fluid give a fast response but little protection against corrosion. The sensors in fig. 2.6 are totally enclosed, but the increased mass gives longer time constants.

2.4.3. Practical circuits

An RTD exhibits a change in resistance with temperature. Before it can be used for measurement or control, this change in resistance must be converted to a change in voltage or current. The electrical power dissipated in the RTD for this conversion

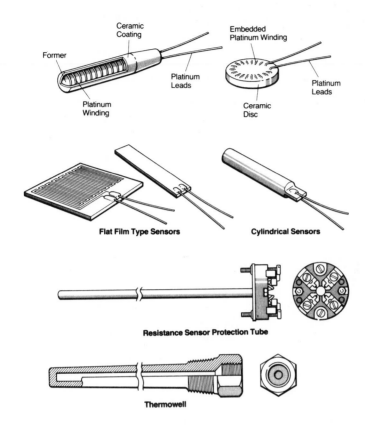

Fig. 2.6 Platinum resistance thermometer construction (courtesy TC Ltd).

must be strictly limited to avoid errors due to I^2R heating of the sensor. Typically 10 mW dissipation will cause a temperature rise of 0.3 °C, which implies low values of current (less than 10 mA) and voltage (below 1 V).

The simplest circuit, shown in fig. 2.7, uses a constant current source to convert the resistance change to a voltage change, V_T, where:

$$V_T = IR_0 (1 + \alpha T) \tag{2.8}$$

IC1 is a unity gain differential amplifier, with V_r corresponding to IR_0. The output voltage is then proportional to T °C.

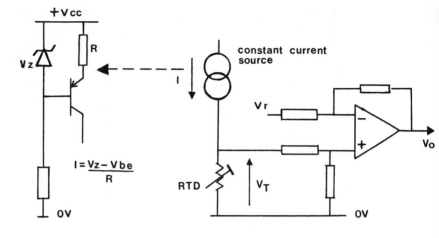

Fig. 2.7 Resistance thermometer circuit.

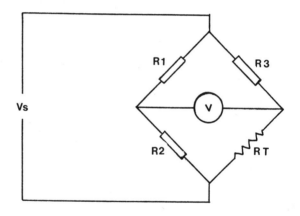

Fig. 2.8 Wheatstone bidge.

The commonest circuits, however, are based on the Wheatstone bridge of fig. 2.8. If the measuring circuit has a high impedance (so that it does not load the bridge), simple circuit analysis shows that:

$$V = V_s \left(\frac{R_T}{R_T + R_3} - \frac{R_2}{R_1 + R_2} \right) \tag{2.9}$$

Unfortunately, because R_T appears in both the numerator and denominator of the left-hand term, *V* does not change linearly with changes in R_T. There are three common ways of overcoming this non-linearity, shown in fig. 2.9.

The non-linearity can be reduced to acceptable levels by making $R_3 >> R_T$ and $R_1 >> R_2$ (typically by a factor of 100). This has the side-effect of reducing the bridge voltage by a factor of 100 as well, but this can easily be re-established by means of a DC amplifier.

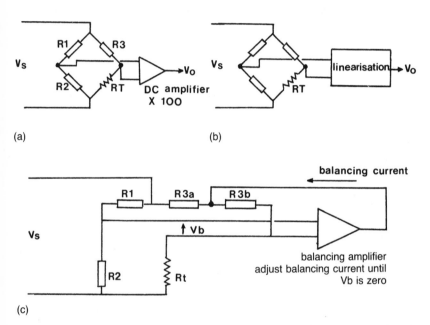

(a)   (b)

(c)

Fig. 2.9 Methods of overcoming non-linearity of bridge output signal. (a) Making R3>>RT and R1>>R2. (b) Linearisation of bridge output. (c) Null balancing technique.

Non-linearity does not, in itself, imply inaccuracy. Orifice plates, for example, have a very non-linear response, but are used to measure flow with minimal errors. In fig. 2.9b, the non-linear output from the bridge is processed by a suitable linearising circuit to give an output voltage which is linearly related to temperature. The linearising can be performed by an Op Amp circuit (see chapter 1 in Volume 2) or by a microprocessor 'intelligent' instrument.

Figure 2.9c uses an electronic circuit to balance the bridge. This is a modern version of the original Wheatstone bridge which was nulled by adjusting the resistance in one of the arms; the desired parameter being inferred by the new resistor value. The null circuit measures the out-of-balance voltage V_b, and increases, or decreases, the current injected into the one arm of the bridge to bring V_b to zero. With the bridge balanced, the change in R_T is simply obtained from the new value of I.

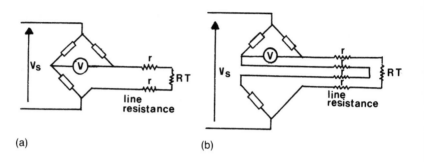

(a)   (b)

Fig. 2.10 The effects of line resistance on RTDs. (a) Two-wire circuit. (b) Four-wire circuit.

In most industrial applications, the RTD will be situated remote from its measurement electronics. If the connecting leads are more than a few metres in length, they will introduce an unknown resistance, r, into each lead, as shown in fig. 2.10a. This unknown resistance is itself subject to change caused by temperature and strain effects, and is a possible source of error. This can be overcome by using a four-wire connection to the RTD, as shown on fig. 2.10b. Each lead will experience the same conditions, so the changes introduced into the RTD leads will be matched by the changes in the dummy leads. The bridge output will only be dependent on changes in temperature of the RTD, not the leads.

In many applications, the three-wire connections of fig. 2.11 give adequate compensation with a small cost saving. Figure 2.11a is the commonest industrial circuit for RTDs.

2.4.4. Thermistors

RTDs utilise the small, but essentially linear, increase in resistance of metals with increasing temperature. Semiconductor

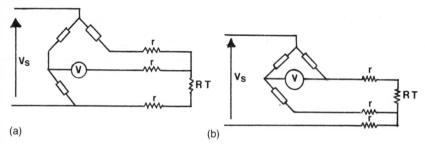

Fig. 2.11 Three-wire connection commonly used in industry. (a) First arrangement. (b) Second arrangement.

materials, however, exhibit a large, but very non-linear, decrease in resistance with increasing temperature. Temperature sensors based on semiconductors are called thermistors.

The variation of resistance for a typical thermistor is shown in fig. 2.12. The resistance of this device decreases from over 10 k-ohm at 0 °C to less than 200 ohm at 100 °C, a change of 50 : 1. The figure also shows the non-linearity inherent in thermistors.

The resistance of a thermistor is defined by:

$$R = A\exp\left(\frac{B}{T}\right) \tag{2.10}$$

where A and B are constants for the particular thermistor, and T is the temperature in degrees Kelvin. The constant B, which has the dimensions of temperature, is called the characteristic temperature and is typically between 3000 K and 5000 K.

Equation 2.10 is not particularly useful, as it is not easy to see what value of resistance will be obtained in the more practical temperature range of $-100°C$ to $+200°C$. Data sheets usually define the resistance R_0 at some temperature T_0 (often 0 °C, 273 K). The resistance at any other temperature is then:

$$R = R_0 \exp B \left(\frac{1}{T} - \frac{1}{T_0}\right) \tag{2.11}$$

where B is the characteristic temperature. Note that T and T_0 are both in degrees Kelvin. The device of fig. 2.12, for example, has a resistance at 25 °C of 3 k-ohm and a value for B of 4020 K.

Thermistors come in a wide variety of shapes, sizes and enclosures, some of which are shown in fig. 2.13. Most are considerably smaller than RTDs, and consequently have a faster

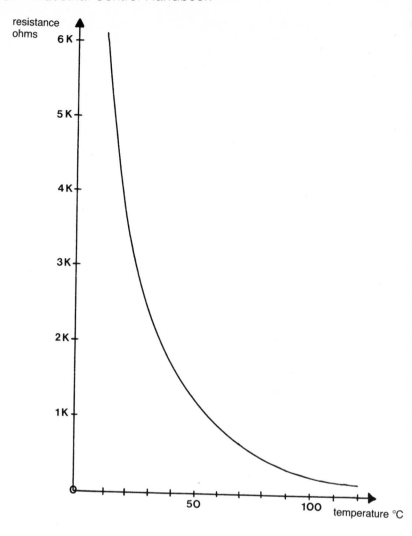

Fig. 2.12   Resistance/temperature curve for a NTC thermistor.

response. Although the response of thermistors is non-linear, they can be used for temperature measurement over a limited range (say 0 °C to 100 °C). A typical circuit, shown in fig. 2.14, produces a voltage at point A given by:

$$V_A = \frac{R_1}{R_1 + R_T} \cdot V_{cc} \tag{2.12}$$

This gives a non-linear response to changes in R_T which roughly

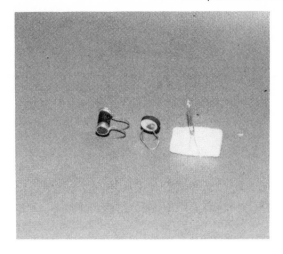

Fig. 2.13 Examples of thermistor types.

compensates for the non-linearity of the thermistor. RV_1 sets a voltage at its slider equivalent to V_A at 0 °C. RV_2 sets the full-scale current through the meter.

The sensitivity of thermistors makes them ideally suited for temperature alarm circuits. In these applications, where all that is required is a signal that a temperature has gone above (or below) some preset, the non-linearity is of little importance.

The thermistors described so far are known as NTC thermistors for negative temperature coefficient. It is also possible to manufacture PTC (positive temperature coefficient) thermistors

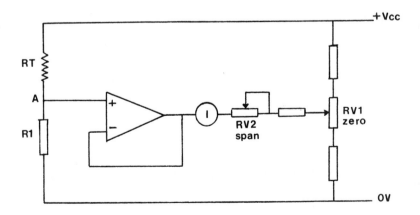

Fig. 2.14 Thermistor-based thermometer.

which exhibit a non-linear increase in resistance with temperature. These have the response of fig. 2.15, which shows that the resistance increases suddenly at some temperature (called the reference temperature). The response of PTC thermistors makes them unsuitable for temperature measurement, but they are widely used as temperature alarm devices for, say, motor windings.

resistance ohms

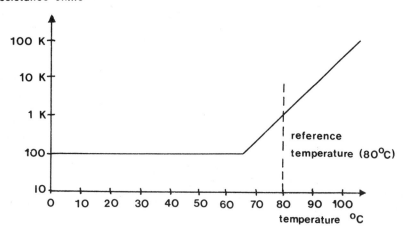

Fig. 2.15   Change in resistance of a PTC thermistor.

2.5.   Thermocouples

2.5.1.   *The thermoelectric effect*

In fig. 2.16, two dissimilar metals are joined at two points as shown. If one end is heated to a temperature T_1, and the other end kept at a lower temperature T_2, a current will flow around the circuit. The current depends on the metals and the temperatures T_1, T_2. This phenomenon, discovered by the Victorian scientist Seebeck, is called the thermoelectric (or Seebeck) effect, and can be used as an accurate measurement of temperature. Devices using this effect are called thermocouples.

The effect arises because an electrical potential arises across the junction of two dissimilar metals. This potential depends on the temperature of the junction, and occurs because of different

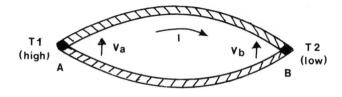

Fig. 2.16 Simple thermocouple.

electrical and thermal properties of the metals. Somewhat simplified, electrons at a higher temperature T_1 have more thermal energy than those at the cooler end in each metal, and there is a drift of electrons towards T_2. The difference in this drift between the two metals produces the voltage. The potential is small, typically a few tens of millivolts.

At junction A, there is a voltage V_a which is a function of T_1. Similarly, at junction B there is a voltage V_b which is a function of T_2. The current arises because V_a differs from V_b. Obviously if $T_1 = T_2$ no current will flow. Inherently, therefore, a thermocouple is a *differential* temperature-measuring device.

2.5.2. Thermocouple laws

The current flowing in the circuit, as in fig. 2.16, is not a convenient indication of temperature, as it depends on the size and length of the wires. The potentials across the junctions are, however, independent of the wire size or length, being solely determined by the metals and the temperature. Introducing a voltmeter into the circuit can introduce errors, however, as new dissimilar metal junctions will be formed in the circuit.

The effects of introducing measuring instruments into the circuit (and other important considerations) are described by five thermocouple laws, illustrated in fig. 2.17.

Law 1 states that the thermoelectric effect depends only on the temperatures of the junctions, and is unaffected by intermediate temperatures along the wires. In fig. 2.17a, the thermocouple wires pass through an area at temperature T_3. The thermocouple effect in this circuit, however, still only depends on T_1 and T_2. This is extremely important in practical locations where the temperature of connecting leads is not known.

Law 2 allows additional metals to be introduced in the circuit without affecting the potentials *provided* junctions of each metal are at the same temperature. In fig. 2.17b new metals are added with junctions CD and EF. These will not affect the circuit provided $T_c = T_d$ and $T_e = T_f$. At each junction, a contact potential will exist, but will be equal and opposite (and hence cancel) if the junction temperatures are the same. Thermocouple cables can be run through connectors, terminal strips and such devices without error, provided temperature differences do not occur across the device.

The third law is an extension of law 2, and states that a third metal can be introduced at either junction, as in fig. 2.17c, without effect, provided that both junctions of the third metal (T_c, T_d) are the same. This has obvious practical implications, as it allows mechanically strong junctions to be made by using brazed, welded or soldered joints. Figure 2.17c also represents the commonest measuring technique, with the third wire being the millivoltmeter.

Law 4, illustrated in fig. 2.17d, is called the law of intermediate metals, and can be used to determine the voltage of, say, a thermocouple based on iron/copper given tables for constantan/copper and iron/constantan.

The final law, called the law of intermediate temperature, is of particular importance when interpolating thermocouple tables. The contact potentials across the junctions are dependent on *absolute* (degrees Kelvin) temperature, and have the form:

$$E_T = AT + BT^2 + CT^3 + \ldots \tag{2.13}$$

where A, B, C, etc. are constants, and T is the temperature. For most applications, terms above the square are ignored.

The thermocouple is a temperature differential measuring device. If T_1, T_2 are the temperatures of the junctions, the resultant voltage is given by:

$$
\begin{aligned}
E_{T_1 T_2} &= E_{T_1} - E_{T_2} \\
&= A(T_1 - T_2) + B(T_1^2 - T_2^2) \tag{2.14}
\end{aligned}
$$

This is a non-linear response, so the same temperature difference does not result in the same voltage. For example, if a type R thermocouple is used, and T_1 is 1250 °C, T_2 is 50°C, the resulting voltage is 13.626 mV. If T_1 is reduced to 1200 °C and T_2 to 0 °C (maintaining the differential of 1200 °C), the voltage changes to

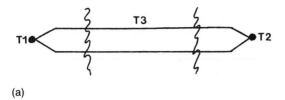

(a)

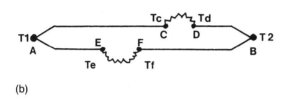

(b)

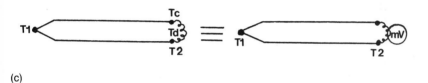

(c)

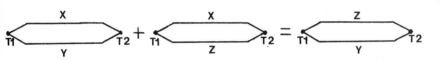

(d)

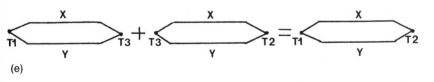

(e)

Fig. 2.17 The five thermocouple laws. (a) First thermocouple law. (b) Second thermocouple law. (c) Third thermocouple law. (d) Fourth thermocouple law (intermediate metals). (e) Fifth thermocouple law (intermediate temperatures).

13.224 mV, a difference of 3% (and a temperature error of about 25 °C).

Tables of thermocouple voltages are published for various temperatures, measured as fig. 2.17c, with T_2 at some reference voltage (usually 0 °C or 20 °C). The law of intermediate

Table 2.1 Thermocouple types

Type	Material +	Material −	$\Delta V/°C$ at 100°C (μV)	Usable range (°C)	Comments
E	Chromel (90% nickel, 10% chromium)	Constantan (57% copper, 43% nickel)	68	0 to 800	Highest-output thermocouple
T	Copper	Constantan	46	−185 to +300	Used for cryogenics and mildly oxidising or reducing atmospheres (e.g. boiler flues)
K	Chromel	Alumel (94% nickel, 3% manganese, 2% aluminium, 1% silicon)	42	0 to 1100	General purpose, widely used
J	Iron	Constantan	46	20 to 700	Used with reducing atmospheres. Iron tends to rust and oxidise; can be improved with chrome/nickel/titanium steel
R	Platinum/13% rhodium	Platinum	8	0 to 1600	High temperatures (e.g. steel making). Used in UK in preference to type S
S	Platinum/10% rhodium	Platinum	8	0 to 1600	As type R, but used outside UK
V	Copper	Copper/nickel	—	—	Compensating cable for type K to 80°C can also be used for type T
U	Copper	Copper/nickel	—	—	Compensating cable for types R and S to 50°C

temperatures allow these tables to be used to deduce the voltage for any other values of T_1, T_2. Suppose we have $T_1 = 1100\ °C$ and $T_2 = 40\ °C$ with a type R thermocouple. The tables (referenced to 0 °C) give a voltage of 11.846 mV for 1100 °C and 0.232 mV for 40 °C. The resultant voltage is 11.614 mV. Note that it is *incorrect* to say $1100 - 40\ °C = 1060\ °C$, and read the value for 1060 °C (11.304 mV) from the table. Section 2.5.7 will deal with using tables.

2.5.3. Thermocouple types

Although almost any pair of dissimilar metals can be used to make a thermocouple, over the years various standards have evolved. These have well documented voltage/temperature relationships, and their use gives interchangeability between different manufacturers. These are shown in table 2.1. Figure 2.18 compares thermocouple outputs and shows their useful ranges.

All that is required to manufacture a thermocouple is to joint the requisite materials. In practice, of course, the thermocouple junction needs protection and rigidity. These are provided by a sheath, usually constructed from magnesium oxide to give a fast

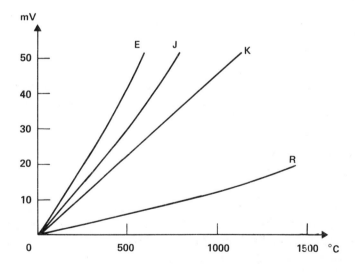

Fig. 2.18 Sensitivity of various thermocouple types.

thermal response. Such assemblies are called mineral insulated metal sheathed thermocouples. The actual junction can be arranged in one of the options of fig. 2.19. An insulated junction gives an electrically isolated output (with an impedance to ground in excess of 100 M-ohm) and total protection from the measured atmosphere. A grounded junction gives a fast thermal response. The fastest response is obtained with an exposed junction, but this can only be used where the atmosphere does not attack the thermocouple wires.

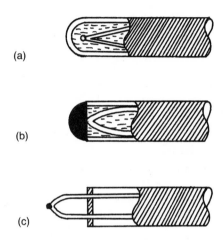

(a)

(b)

(c)

Fig. 2.19 Types of measuring junction configuration. (a) Insulated junction. (b) Grounded junction. (c) Exposed junction.

Thermocouples are available in a variety of sizes and shapes, some of which are shown in fig. 2.20. They can be protected by an enclosure called a thermowell, but this obviously introduces a large lag.

2.5.4. Extension and compensating cables

Equation 2.13 emphasises that a thermocouple is a differential measuring device, and that to be useful one temperature (usually the lower) must be known. This can cause measurement problems if due care is not taken with the installation.

Figure 2.21 shows a typical thermocouple installation where a probe is used to measure the temperature in an oven. In fig.

Type 3. General purpose probe

Suitable for general purpose applications up to 400°C these assemblies are available in thermocouple conductor calibration codes: K, T, J and E as simplex and duplex units.
Type 3 assemblies are supplied as standard with seamless welded closed end sheaths in AISI type 316 stainless steel. Other sheath materials are available from our range of stainless steels, Inconel*, Incoloy* and other alloys.
If the limiting operating temperature of 400°C at the tip of these assemblies is too low then our type 12 assemblies may be more suitable. SEE PAGES 24 & 25.
Junctions are supplied grounded to the sheath as standard with insulated junctions available as an option.

Type No:	Lead arrangement
3AX	PVC leads
3AY	Teflon leads
3AZ	Fibreglass leads
3AS	Stainless steel braid over fibreglass leads
3AF	Galvanised steel conduit over fibreglass leads
3AG	Stainless steel conduit over fibreglass leads
3H11	Thermocouple plug termination on sheath
3M11	Miniature thermocouple plug termination on sheath
3TH	Screw top weatherproof head (3P10) on sheath end

Dimension A (sheath length):
to suit application
Dimension B (lead length):
to suit application
Dimension d (sheath diameter):
inches: ³⁄₁₆, ⅛, ¼
mm: 1.5, 3.0, 3.0, 5.0 or 6.0

Optional extras.
Insulated junction.   Pin or spade termination.
Reduced tip.   120 degree ground tip.
90 degree ground tip.

Moving surface thermocouple

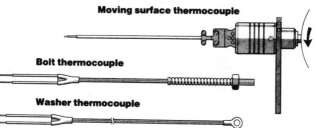

Bolt thermocouple

Washer thermocouple

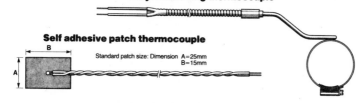

Adjustable ring thermocouple

Self adhesive patch thermocouple

Standard patch size: Dimension A=25mm
B=15mm

Hand held thermocouple probe

High temperature industrial ceramic sheathed thermocouples

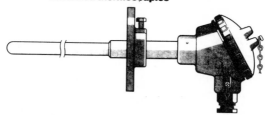

Fig. 2.20   A selection of thermocouple types (courtesy TC Ltd).

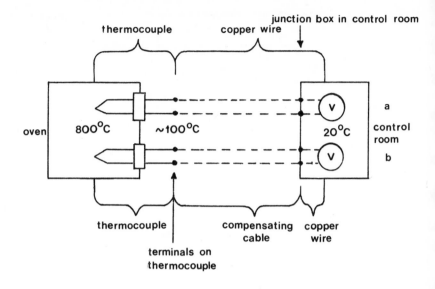

Fig. 2.21 The use of compensating cable.

2.21a, the probe terminals are connected back to the control room by ordinary copper wires. By law 3, and fig. 2.17c, the voltage seen at the control room will be determined by the oven temperature, and the *terminal* temperature of the thermocouple probe. The latter is unknown, but will probably be high due to heat conducted out of the oven, and will vary from day to day.

An improved connection, fig. 2.21b, uses matching leads of the same material as the thermocouple to connect the probe to the control room. The voltage now seen is a function of the oven temperature and the control-room temperature. The latter will be around 20 °C, and will be known and relatively stable. Meaningful measurements can now be made.

The connection leads do not have to be made to as tight a standard as the thermocouples, and are known as extension cables. With type R and S thermocouples, however, the cost of extension cables would be prohibitively high and an alternative approach is used. In most applications, the temperature difference between ends of the connection cable is small compared with the temperature being measured (in fig. 2.21, for example, it is about 80 °C). The cost of connecting cables can be reduced considerably by using metals which match the thermocouple over a *limited* temperature range. Such cables are called compensating cables. Type U on table 2.1, for example, is a copper/copper–nickel cable manufactured to match types R and S thermocouples up to 50 °C. If the thermocouple terminals are likely to exceed this, an alternative compensating cable must be used, or extension cables taken to an area below 50 °C.

Compensation/extension cables are colour coded according to type. There are different standards in different countries, as shown in table 2.2, but fortunately there are no ambiguities. A red sheath with brown/blue leads, for example, can only be a type K extension cable to UK standard BS 1843.

It is essential, of course, to maintain the correct polarity from thermocouple, through the compensating/extension cable to the measuring device.

2.5.5. Cold junctions and cold junction compensation

Figure 2.21b allows measurement of the thermocouple temperature, but for accurate results the control-room temperature must be known and some simple arithmetic must be performed on thermocouple tables. The operation is simplified if the junction of the extension cable/measuring instrument (called the cold junction) is kept at a fixed temperature.

A possible approach (called an ice cell) is to use an ice/water mix to keep the junction at 0 °C. This allows 0 °C referenced tables to be used directly. An ice cell is, however, inconvenient for industrial use.

A more convenient solution, called cold junction compensation (shown in fig. 2.22), is to measure the temperature of the cold junction by an RTD or thermistor and correct the thermocouple indicated temperature. This can be done by adding a correction temperature to the instrument, as in fig. 2.22a, or by modifying the thermocouple signal directly, as in fig. 2.22b.

Where thermocouple signals are to be transmitted over long distances, a combined cold junction compensator/amplifier may be used. A typical device is shown in fig. 2.23. This unit accepts a millivolt signal from a type R thermocouple and outputs a 4–20 mA current signal corresponding to 0–1400 °C.

Table 2.2 Colour coding of thermocouple cables

Type	British BS 1843			American ANSI			German DIN 43714			French NFC 42–323		
	Sheath	+	–	Sheath	+	–	Sheath	+	–	Sheath	+	–
E	Brown	Brown	Blue	Violet	Violet	Red	Black	Red	Black	–	–	–
T	Blue	White	Blue	Blue	Blue	Red	Brown	Red	Brown	Blue	Yellow	Blue
K	Red	Brown	Blue	Yellow	Yellow	Red	Green	Red	Green	Yellow	Yellow	Violet
J	Black	Yellow	Blue	Black	White	Red	Blue	Red	Blue	Black	Yellow	Black
R	Green	White	Blue	Green	Black	Red	–	–	–	–	–	–
S	Green	White	Blue	Green	Black	Red	White	Red	White	Green	Yellow	Green
V*	Red	White	Blue	–	–	–	–	–	–	Red	Yellow	Brown
U*	Green	White	Blue	Green	Black	Red	White	Red	White	Green	Yellow	Green

* Compensating cable

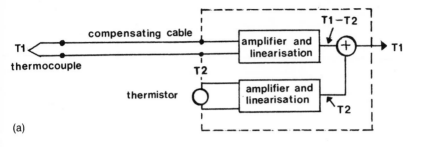

(a)

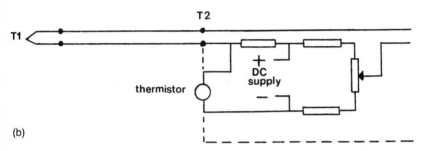

(b)

Fig. 2.22 Cold junction compensation. (a) Addition of cold junction temperature. (b) In-line compensation.

2.5.6. *Thermocouple temperature indicators*

The main problem encountered with thermocouples is the low signal level. Electrical noise on the signal lines will almost certainly be several orders of magnitude higher than the temperature signal itself. The measuring device must therefore have a high input impedance and very high common-mode noise rejection. Particular care needs to be taken with the cable installation with regard to screening and the avoidance of ground loops.

Equation 2.13 shows that linearisation will be needed in all but the least critical applications. This must be performed in the instrument, either via Op Amp diode break circuits (see chapter 2 in Volume 2) in analog instruments, or via look-up tables and direct computation in digital instruments. In most instruments, the cold junction compensation will be included in the instrument itself.

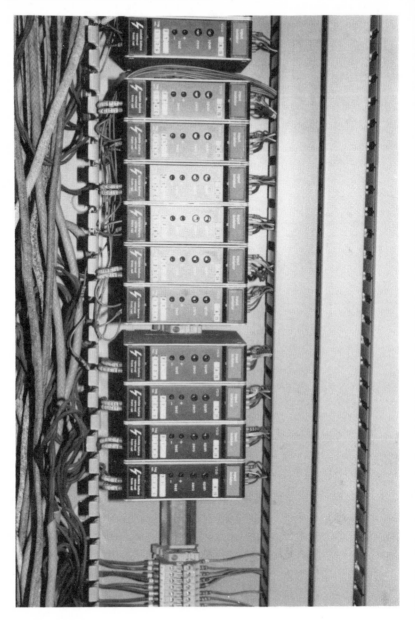

Fig. 2.23 Combined cold junction compensation & isolation amplifiers. Manufactured by TCS Ltd.

2.5.7. Using thermocouple tables

Thermocouple tables are provided by the manufacturers. These detail the voltages obtained at various temperatures. Table 2.3 shows an extract from the table for a type R thermocouple. These tables may be used for maintenance and calibration in two circumstances: for checking the voltages from a thermocouple with a millivoltmeter (fig. 2.24a) or for injecting a test voltage from a millivolt source into a temperature indicator (fig. 2.24b). In each case, the ambient temperature must be known to apply the law of intermediate temperatures. In the examples below, type R tables are used.

Table 2.3 Extract from thermocouple tables for type R (platinum–platinum/ 13% rhodium) (Referenced to cold junction at 0 °C. Voltage in µV.)

	0	1	2	Temperature (°C) 3	4	5	6	7	8	9
0:	0	5	11	16	21	27	32	38	43	49
10:	54	60	65	71	77	82	88	94	100	105
20:	111	117	123	129	135	141	147	152	158	165
800:	7949	7961	7973	7986	7998	8010	8023	8035	8047	8060
810:	8072	8085	8097	8109	8122	8134	8146	8159	8171	8184
820:	8196	8208	8221	8233	8246	8258	8271	8283	8295	8308
830:	8320	8333	8345	8358	8370	8383	8395	8408	8420	8433
840:	8445	8458	8470	8483	8495	8508	8520	8533	8545	8558
850:	8570	8583	9595	8608	8621	8633	8646	8658	8671	8683
930:	9589	9602	9614	9627	9640	9653	9666	9679	9692	9705
940:	9718	9731	9744	9757	9770	9783	9796	9809	9822	9835
950:	9848	9861	9874	9887	9900	9913	9926	9939	9952	9965
960:	9978	9991	10004	10017	10030	10043	10056	10069	10082	10095

For fig. 2.24a, there are four steps:

(1) Measure the ambient temperature (a mercury-in-glass thermometer will suffice). Read the corresponding voltage from the tables. For an ambient of 22 °C, say, the tables give 0.123 mV.

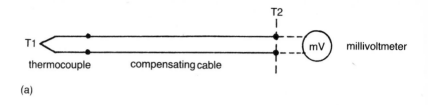

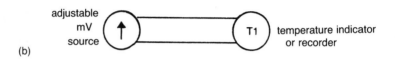

(a)

(b)

Fig. 2.24 Using thermocouple tables. (a) Checking thermocouple output. (b) Checking temperature measuring device.

> (2) Measure the thermocouple voltage. Let us assume this is 9.64 mV.
> (3) Add the ambient voltage: 9.64 + 0.123 = 9.763 mV.
> (4) Find the temperature on the table corresponding to this.

The tables show 9.757 mV at 943 °C, and 9.770 mV at 944 °C, so our temperature is about 943.5 °C. The precise temperature could be found by interpolation, but 1 °C is sufficiently accurate for most purposes.

Note that reading the temperature for 9.64 mV (934 °C) and adding 22 °C (to give 956 °C) givs an error of 13 °C. The non-linearity of equation 2.13 requires the above procedure to be followed.

With fig. 2.24b, the injection voltage required to give a particular temperature indication is to be found.

> (1) Measure the ambient temperature and find the corresponding table voltage; say 18 °C, which gives 0.1 mV.
> (2) Find the table voltage for the test temperature; say 850 °C, which gives 8.57 mV.
> (3) Subtract the ambient voltage from the test temperature to give the required injection voltage (8.57 − 0.1 = 8.47 mV).

Note again that an injection voltage corresponding to 850 °C − 18 °C = 832 °C (8.345 mV) will give an incorrect result. In this case the instrument would display 840 °C, an error of 10 °C.

2.6. Radiation pyrometry

2.6.1. Introduction

When an object is heated, it radiates electromagnetic energy. At low temperatures this radiation can be felt; as the temperature rises it starts to emit visible radiation (i.e. light), passing from red heat, through yellow to white heat. Intuitively, this radiation can be used to measure temperature; qualitatively we can say an object glowing yellow is hotter than an object glowing dull red. Pyrometers use the same radiation to measure temperature.

Pyrometers allow non-contact measurement of temperature, which is essential where the temperature of a moving object is to be measured, or where an environment exists that would destroy a more conventional sensor.

Light, radio waves, X-rays, infrared and ultraviolet are all electromagnetic waves and part of the electromagnetic spectrum (fig. 2.25). The difference between them is simply their frequency, which ranges from below 10^4 Hz for radio waves, through 10^{14} Hz for visible light, to over 10^{20} Hz for gamma and X-rays. Electromagnetic radiation can also be described by its wavelength, λ, which is related to the frequency, f, by:

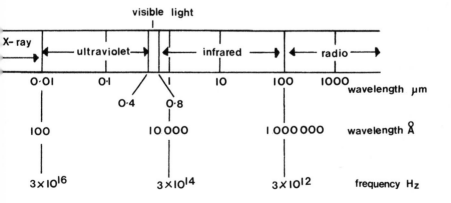

Fig. 2.25   The electromagnetic spectrum.

$$C = f\lambda \qquad\qquad (2.15)$$

where C is the speed of light (3×10^{10} cm/sec).

Although we are only physically aware of radiated energy when an object achieves a temperature of about 200 °C, *all* objects at a temperature above absolute zero (0 K) emit radiation whose power and frequency distribution are temperature dependent.

2.6.2.   Black-body radiation

Any object is constantly receiving energy from its surroundings. Figure 2.26 shows an object receiving energy W; of this a proportion εW is absorbed (and a proportion $(1 - \varepsilon)W$ reflected), where ε is defined as the emissivity of the object. If the object absorbs all the incident energy (i.e. $\varepsilon = 1$), it is called a black body.

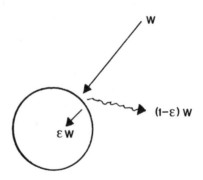

Fig. 2.26   Definition of emissivity.

The term is actually a misleading one, as it implies the object is 'black' in colour. A black body, like all objects, is also an emitter of radiation and when sufficiently raised in temperature will emit visible light. The sun, for example, is an almost perfect black body with a temperature of 6000 K; a hole in the side of an enclosed oven is a good approximation to a black body.

Any object at a steady temperature and not subject to convection or conduction losses must be radiating the same amount of energy as it receives (otherwise there would be a net gain or loss of energy and the temperature would rise or fall). The energy radiated is not at one specific frequency, but is spread over a range of frequencies. The relative radiated power density of a black body is plotted against wavelength for various temperatures on fig. 2.27. Note that logarithmic scales are used for both power density and wavelength axis. Pyrometry is mainly used in the range 0.3 to 100 μm, which approximately corresponds to the visible light and infrared regions.

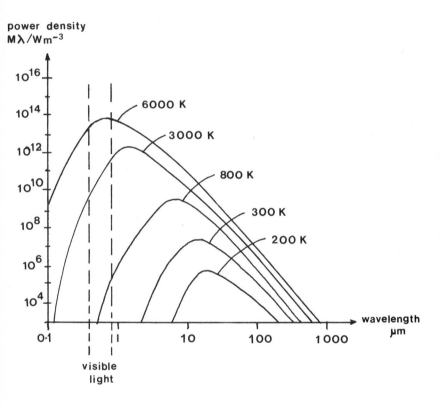

Fig. 2.27 Relative radiated power density for a black body at different temperatures.

Figure 2.27 shows that the total radiated power (which is the area under any curve) increases dramatically with temperature. The total power at any temperature is given by Stefan's law, which states that:

$$M = \sigma T^4 \quad \text{W m}^{-2} \tag{2.16}$$

where σ is Stefan's constant (5.67×10^{-8} W m^{-2} K^{-4}) and T is measured in Kelvin.

As well as increasing in magnitude, the curves peak at shorter wavelengths (higher frequencies) for increasing temperature. At around 800 K, there is a significant amount of power radiated in the visible region of the spectrum, and we see the object start to glow.

Very few real-life objects are black bodies: most reflect radiation to some degree. If a real-life object absorbs less radiation than a black body, it must also radiate less (or there will be a net loss of energy and the object's temperature will fall). This is expressed as Kirchhoff's law:

> If a body at a certain temperature T absorbs a certain fraction ε of the radiation of wavelength λ inclined upon it, then it will emit the same fraction ε of the radiation at that wavelength that a black body of the same temperature T would emit.

This can be expressed more succinctly by:

$$\varepsilon = \frac{\text{radiation emitted by an object at temperature } T}{\text{radiation emitted by black body at temperature } T}$$

at any wavelength λ. The value of ε varies with the wavelength of the radiation. Refractory brick, for example, has an emissivity which varies as shown in fig. 2.28. Obviously the emissivity

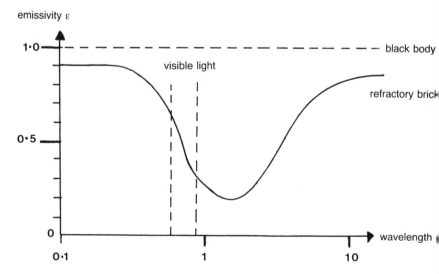

Fig. 2.28   Emissivity of refractory brick.

introduces a large potential error into pyrometry.

2.6.3. Principles of pyrometry

A pyrometer is, in theory, a very simple device, as shown on fig. 2.29a. The object whose temperature is to be measured is 'viewed' through a fixed aperture by a temperature measuring device. Part of the radiation emitted by the object falls on the temperature sensor, and causes its temperature to rise. The object's temperature is deduced from the sensor temperature. The temperature sensor itself must be of small thermal mass to give reasonable sensitivity. Usually a circular ring of thermocouples in series (called a thermopile) is arranged as shown in fig. 2.29b, with a diameter of about 2–10 mm. Alternatively a small resistance thermometer (called a bolometer) may be used.

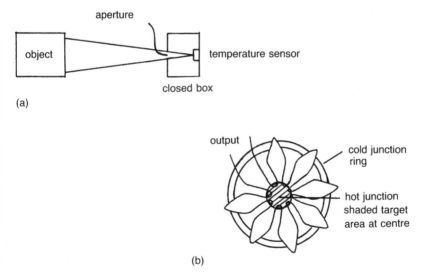

Fig. 2.29 Optical pyrometer. (a) Principle of optical pyrometry. (b) Thermopile.

One of the advantages of pyrometers is that the measurement is independent of the distance from the object (provided the object fills the field of view). Any point on the surface of the object radiates in all directions. The amount of radiated energy received by the sensor will be proportional to the solid angle subtended by the sensor, as shown in fig. 2.30a. Less energy will be received, from a given point, at position B than at position A. The angle varies inversely as the square of the distance from the object to the sensor.

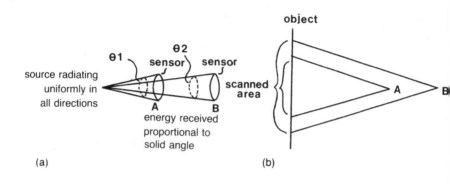

Fig. 2.30 Pyrometry and distance from sensor to object. (a) Energy received from given point decreases with distance. (b) Scanned area increases with distance.

There is another factor at work, however. As the sensor moves further away, the surface area scanned increases, as shown on fig. 2.30b, so the total radiation being scanned increases with distance. Simple geometry shows that the area scanned increases as the square of the distance.

These two effects cancel, and the radiation received by the sensor remains constant with distance. In practice, the object must fill the field of view, and absorption of radiation by air causes a slight fall-off with distance.

The design of a practical instrument is shown in fig. 2.31. This is a hand-held device, and is aimed at the target object by means of an eyepiece at the rear of the instrument. A concave mirror focuses the radiation on to the thermopile, which is supported on the centreline of the device. The mirror focusing reduces the overall length of the instrument, and increases the sensitivity. The thermopile output is converted to a temperature indication by electronics in the handle. The meter is viewed through the eyepiece, and is superimposed on the field of view by suitable optics.

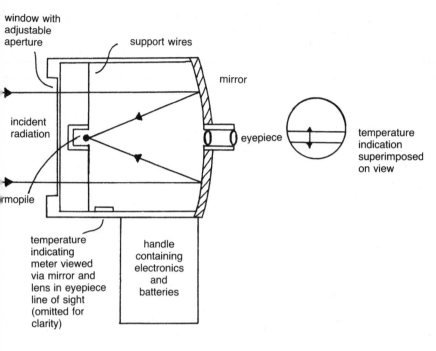

Fig. 2.31 Hand-held optical pyrometer.

The output from the thermopile is obviously small and very non-linear. It is also prone to errors due to drifts in the electronic amplifier and linearisation circuit. The latter error can be overcome by the arrangement of fig. 2.32. The detector is housed in a temperature-controlled oven, and the incoming radiation is chopped by a rotating disc. The detector has a very small thermal time constant, so its temperature rises and falls and produces an

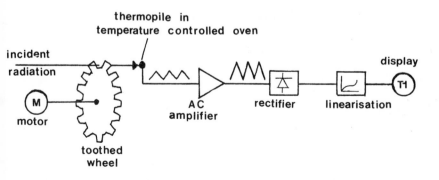

Fig. 2.32 Chopper pyrometer.

alternating voltage as shown. This can be amplified by a drift-free AC amplifier. The resulting high-level signal can then be rectified to give a DC signal and linearised for display.

2.6.4. Types of pyrometer

Although all pyrometers operate on radiated energy, there are various ways in which the temperature is deduced from the radiation absorbed by the sensor. The simplest method is to measure the total power of the radiation, effectively the total area under any curve on fig. 2.27. The unit of fig. 2.31 is of this type.

The curves of fig. 2.27 also show that the power radiated at any frequency also increases with temperature. Monochromatic radiation pyrometers use an optical filter to restrict the measurement to a narrow frequency band. Although this reduces the sensitivity, it allows the choice of a frequency at which the emissivity approaches unity.

Lack of knowledge of emissivity is the major source of error in pyrometry. This can be overcome by a two-colour pyrometer which compares radiation at two frequencies. Examination of fig. 2.27 will show that the ratio of the power at wavelengths of, say, 3 and 4 μm is temperature dependent. Since both frequencies have similar emissivities, the ratio will be independent of the actual emissivity value. This can be achieved by the arrangement of fig. 2.33. The incoming radiation is split equally to pass through two filters to two detectors. The outputs of the two detectors are ratioed electronically.

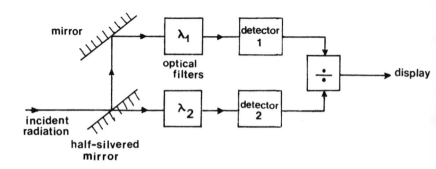

Fig. 2.33 Overcoming unknown emissivity by measurements at different wavelengths.

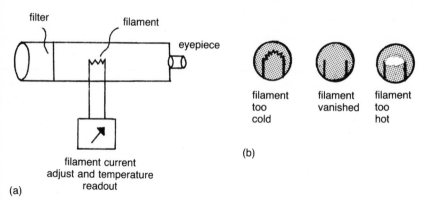

Fig. 2.34 Disappearing filament pyrometer. (a) Principle of instrument. (b) View through eyepiece.

An optical pyrometer, shown in fig. 2.34a, does not use a temperature detector as such. The target object is viewed though a telescope optical system, and a hot wire filament is superimposed on the field of view. The current through the filament, and hence its temperature, can be controlled by the operator. The current is adjusted until the filament merges into the background, as shown in fig. 2.34b. At this point the brightness of the filament is the same as the target object, and the temperature can be deduced from the filament current.

Typical pyrometers are shown in figs 2.35 and 2.36. The former is a hand-held unit with a range of 400–2000 °C and an accuracy of 2 °C. The fixed unit in fig. 2.36 is measuring the wall temperature of a gas-fired kiln; it covers the range 750–1300 °C and has an accuracy of 5 °C, mainly determined by knowledge of the emissivity.

Fig. 2.35 Hand held pyrometer.

Fig. 2.36 Fixed pyrometer. Top pipe is air purge to cool unit & clear dirt from lens.

Chapter 3
Pressure transducers

3.1. Introduction

3.1.1. Definition of terms

Pressure is a very loose and general term. It is scientifically defined as the force per unit area that a fluid exerts on the walls of the container that holds it. For measurement purposes, however, it is necessary to define the conditions under which the measurement is made.

All pure pressure transducers are concerned with the measurement of *static pressure*, i.e. the pressure of a fluid at rest. If the fluid is in motion, its pressure will depend on its flow velocity and is termed the *dynamic pressure*. Section 5.2 describes flowmeters which use this principle. Although an orifice/differential pressure transmitter may be used to measure flow from the fluid's dynamic pressure, the pressure transmitter itself is measuring the static pressure difference between the two pipes from the orifice tapping.

Differential pressure (fig. 3.1a) is the difference between two pressures applied to the transducer. Differential pressure transducers (commonly called Delta P transmitters) are widely used in flow measurement. Conceptually, the differential pressure transducer can be considered the 'basic' pressure measurement device, other types being variations.

If, for example, one pressure input to a differential pressure transducer is left open, the device indicates *gauge pressure* (fig. 3.1b), which is pressure related to atmospheric pressure. Atmospheric pressure can vary, however, so if a more accurate measurement is required, the pressure measurement can be made with respect to a fixed pressure in a sealed chamber. Such a

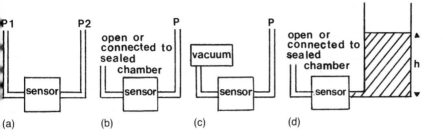

Fig. 3.1 Definitions of 'pressure'. (a) Differential. (b) Gauge. (c) Absolute. (d) Head.

device is called a *sealed gauge pressure transducer.* In UK engineering terms, the suffix 'g' is added to indicate gauge pressure, e.g. 37 p.s.i.g. *Absolute pressure* (fig. 3.1c) is measured with respect to a vacuum. This is effectively a differential pressure measurement with one pressure at zero. *Head pressure* (fig. 3.1d) is a term used in liquid level measurement, and refers to pressure in terms of the height of the liquid (e.g. inches water gauge, millimetres mercury). It is effectively a gauge pressure measurement, but changes in atmospheric pressure affect both the liquid and the pressure-measuring devices equally, provided the liquid is held in a vented vessel.

Vacuum measurement is a form of pressure measurement, but one requiring special devices and techniques. Vacuum transducers are therefore described separately in section 3.8.

3.1.2. Units

There is a wide and confusing range of units for pressure measurement. Scientifically, pressure is force per unit area. This is described in the SI system as newtons per square metre $(N\ m^{-2})$, which has been named the pascal (Pa). This is not a widely used unit, the kilopascal (kPa) or the 'non-standard' $N\ cm^{-2}$ being more convenient. In the UK, pounds per square inch (p.s.i.) is widely used. In many applications, the *atmosphere* (14.7 p.s.i.) is used, or the *bar* (100 kPa), which approximates to an atmosphere. Measurements in inches water gauge or millimetres mercury are also common. Table 3.1 gives conversions between common units.

Table 3.1   Common pressure conversions

1 p.s.i. (lbf/in^2)	=	6.895 kPa
	=	27.7 inches WG
1 lbf/ft^2	=	47.88 Pa
1 kgf/cm^2	=	98.07 kPa
1 inch WG	=	249.0 Pa
	=	5.2 lbf/ft^2
	=	0.036 p.s.i.
1 foot WG	=	2.989 kPa
	=	62.43 lbf/ft^2
	=	0.433 p.s.i.
1 torr (mmHg)	=	133.3 Pa
1 bar	=	100 kPa
	=	14.5 p.s.i.
	=	750 mmHg
	=	401.8 inches WG
	=	1.0197 kgf/cm^2
1 atmosphere	=	1.013 bar
	=	14.7 p.s.i.
1 kilopascal	=	0.145 p.s.i.
	=	20.89 lbf/ft^2
	=	1.0197×10^{-3} kgf/cm^2
	=	4.141 inches WG
	=	7.502 torr (mmHg)
	=	0.01 bar
	=	9.872×10^{-3} atm

3.2.   U-tube manometers

Although manometers are not widely used nowadays in industry, they give a useful insight into the principle of pressure measurement. They are also used as a standard against which other devices may be calibrated.

If a U-tube is filled with a liquid (commonly water, alcohol or mercury, depending on the pressure to be measured), the liquids will naturally adopt the same height in both tubes, as shown in

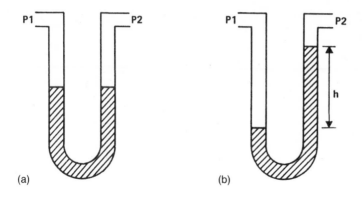

Fig. 3.2 The U-tube manometer. (a) P1 = P2. (b) P1>P2.

fig. 3.2a. If a pressure is applied to each leg, as shown in fig. 3.2b, the liquid level will fall on the high-pressure side and rise on the low-pressure side to give a height difference, h.

The head pressure of a column of liquid is given by:

$$p = \varrho g h \tag{3.1}$$

where p is the pressure (Pa), ϱ is the density, g is the acceleration due to gravity (9.8 m s^{-2}), and h is the column height in metres.

In the UK, weight density is used, so equation 5.1 becomes

$$p = \varrho h \tag{3.2}$$

where p is in pounds per square foot, ϱ is in pounds per cubic foot and h is in feet.

In fig. 3.2b, the pressures in both tubes must balance, so

$$P_1 = P_2 + \varrho g h \tag{3.3}$$

or

$$h = (P_1 - P_2)/\varrho g \tag{3.4}$$

i.e. the difference in column heights is proportional to the differential pressure.

Manometers are usually scaled with a datum at zero, and 'half scaled' divisions above and below the datum (e.g. a device measuring inches water gauge would have divisions spaced 0.5 in apart, but labelled 1 in, 2 in, etc.). This allows the pressure to be read off one tube.

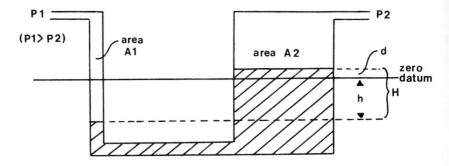

Fig. 3.3 Scaled (unequal area) manometer.

Scaled manometers can be constructed by having unequal areas in the two tubes, as in fig. 3.3. The differential height H still obeys equation 3.4, but the movements from the zero (d and h as shown) are not equal. If A_1, A_2 are the areas of the tubes, the volume displacements in each tube will be equal, so

$$A_1 h = A_2 d \tag{3.5}$$

or

$$h = \frac{A_2}{A_1} d \tag{3.6}$$

since $h + d = H$, substituting in equation 3.4 gives:

$$d = \frac{P_1 - P_2}{\varrho g (1 + A_2/A_1)} \tag{3.7}$$

The scaling of the manometer can be adjusted by varying A_2/A_1. Sometimes 'range tubes' are provided with manometers which are screw-in tubes of different sizes to allow one area to be altered.

An inclined tube manometer (fig. 3.4) gives increased sensitivity for low-pressure measurements. The pressure is read off an inclined measurement scale from the distance d. If the area of the tube A_2 is considerably less than the area of the reservoir, h is given by equation 3.4 and d by simple trigonometry:

$$d = \frac{(P_1 - P_2)}{\varrho g \tan \theta} \tag{3.8}$$

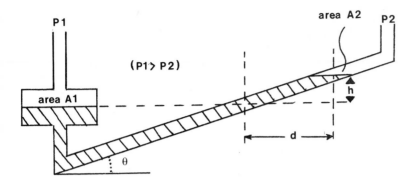

Fig. 3.4 Inclined tube manometer.

3.3. Elastic sensing elements

3.3.1. *The Bourdon tube*

The Bourdon tube, dating from the mid nineteenth century, is still the commonest pressure measuring device where remote indication over a long distance is required or where very high or low pressures are to be measured. The tube itself is manufactured by flattening a circular-cross-section tube to the section shown in fig. 3.5, and bending to a C shape. One end is fixed, left open and connected to the pressure to be measured. The other is closed and left free.

If a pressure is now applied to the inside of the tube, it will tend to straighten, causing the free end of the tube to move up and to the right. This movement is converted to a circular pointer movement by a quadrant-and-pinion mechanical linkage. The movement depends on the pressure difference between the inside and outside of the tube, so the Bourdon tube inherently measures gauge pressure.

Bourdon tubes are usable from 0–5 p.s.i. (0–30 kPa) to about 0–10 000 p.s.i. (0–50 MPa approx.). The lower ranges employ a spiral tube to increase the sensitivity.

Where an electrical output is required, the Bourdon tube can be coupled to a potentiometer giving an electrical output that varies with applied pressure between 0 and 100% of the applied voltage.

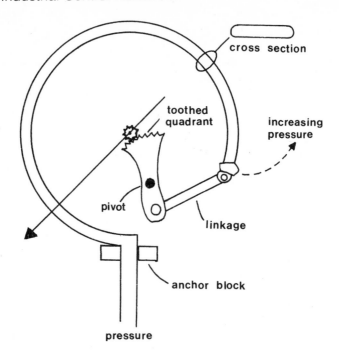

Fig. 3.5 Bourdon tube.

3.3.2. *Various sensing elements*

The Bourdon tube is only one example of an elastic sensing element. Pressure transducers using this principle convert a differential pressure to a displacement which can be measured by any of the methods in sections 3.3.3 and 3.3.4.

Figure 3.6 is a schematic representation of a differential pressure transducer. Input pressures P_1, P_2 ($P_1 > P_2$) are applied to a cylinder containing a movable plate of area A. A force is produced on the piston of $A.(P_1 - P_2)$. The plate is restrained by a spring of stiffness K, so the plate moves until the force exerted by the spring equals the force caused by the pressure differential, i.e.

$$Kd = A.(P_1 - P_2) \tag{3.9}$$

or

$$d = \frac{A.(P_1 - P_2)}{K} \tag{3.10}$$

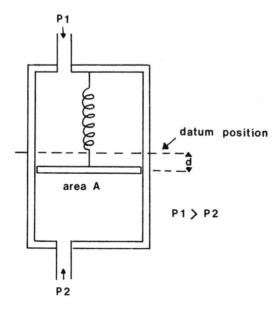

Fig. 3.6 Differential pressure transducer.

The displacement is proportional to the pressure differential.

The arrangement of fig. 3.6 is obviously impractical (the plate edges would be impossible to seal without excessive friction, for example). Figure 3.7 shows common practical elements; all of these convert a pressure differential to a linear displacement. The designs provide a seal between the low- and high-pressure sides, and are constructed to minimise frictional effects and give a linear pressure-to-displacement relationship.

The linear movement can be amplified mechanically to drive a pointer (the common domestic aneroid barometer, for example, uses the capsule of fig. 3.7d) or converted to an electrical signal as described below.

3.3.3. Electrical displacement sensors

The displacement of all the devices in fig. 3.7 is small: at most a few millimetres. For remote indication this displacement must be converted to an electrical signal (pneumatic force balance transducers are described in section 3.5.2). The devices described below are all position measuring devices, and are dealt with in

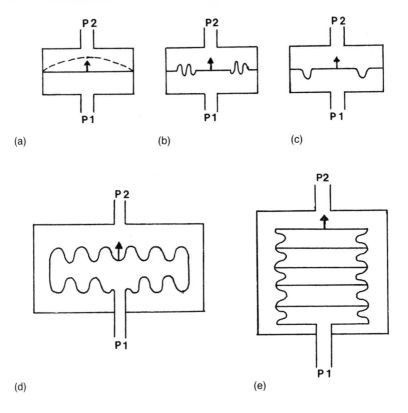

Fig. 3.7 Elastic pressure sensing elements. (a) Diaphragm. (b) Corrugated diaphragm. (c) Catenary diaphragm. (d) Capsule. (e) Bellows.

more detail in section 4.5.

Strain gauges are often used with the simple diaphragm of fig. 3.7a where the displacement is usually small. Care must be taken to provide correct temperature compensation. Often the strain gauge(s) is (are) simply fixed direct to the surface of the diaphragm itself.

Figure 3.8 shows various arrangements of *capacitive sensors*. In fig. 3.8a the diaphragm acts as one plate of a capacitor. As the diaphragm moves with respect to the fixed plate, the change in plate separation causes a change in capacitance. Figure 3.8b uses the same principle, but the movable plate moves between two fixed plates, causing one capacitance to increase and one to decrease. This increases the sensitivity, and is convenient for use with an AC bridge circuit. Figure 3.8c uses the displacement to move one plate at a constant distance from the fixed plate. The

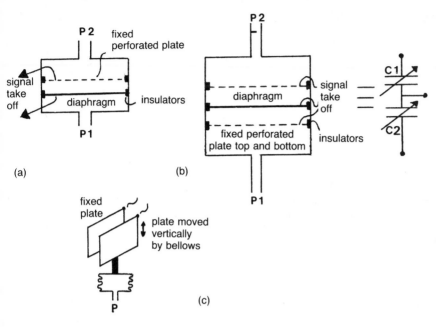

Fig. 3.8 Variable capacitor pressure sensing elements. (a) Single capacitor, variable spacing. (b) Double capacitor, variable spacing. (c) Variable area.

change in area causes a change in capacitance.

The *LVDT (linear variable differential transformer)* of fig. 3.9 uses the displacement from the elastic sensing element to move a ferromagnetic core in a differential transformer. The displacement alters the coupling between the primary (fed typically at a few kilohertz) and the two secondaries, causing the voltages E_1, E_2 to differ. The output voltage magnitude $(E_1 - E_2)$ and phase therefore depend on the core movement. A phase-sensitive rectifier gives a DC output signal proportional to the core displacement from the centre (balanced) position.

The elastic sensing displacement in fig. 3.10 moves a ferromagnetic plate in front on a coil wound on an E core. Movement of the plate causes a change in the inductance of the coil, which can be measured by an AC bridge. This arrangement is called a *variable reluctance* or *variable inductance transducer*.

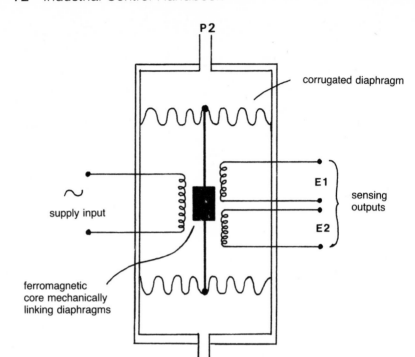

(a)

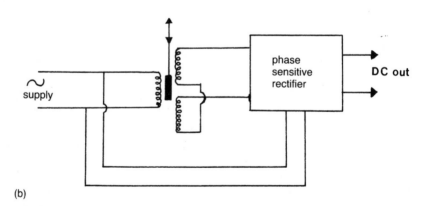

(b)

Fig. 3.9 LVDT pressure transducer. (a) Mechanical arrangement. (b) Electrical circuit.

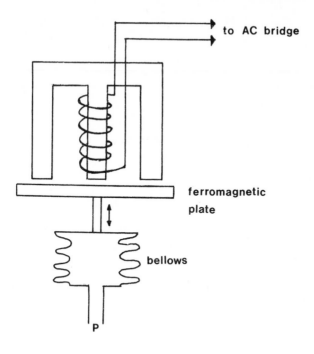

to AC bridge

ferromagnetic
plate

bellows

P

Fig. 3.10 Variable reluctance pressure transducer.

3.4. Piezo elements

The piezo-electric effect occurs in quartz crystals. When a
suitably prepared crystal has a force applied to it (fig. 3.11a),
electrical charges of opposite polarity appear on the faces. The
charge is proportional to the applied force.

To be of use, an output voltage must be produced from the
charge. This is provided by the circuit of fig. 3.11b, which is called
a charge amplifier. C_s is the stray capacitance of the connecting
cable and the crystal itself. Pressure changes across the
diaphragm cause a charge q across the crystal:

$$I_1 = \frac{\mathrm{d}q}{\mathrm{d}t} \tag{3.11}$$

$$I_2 = -C\frac{\mathrm{d}V}{\mathrm{d}t} \tag{3.12}$$

These are equal, so:

$$\frac{dq}{dt} = -C\frac{dV}{dt} \tag{3.13}$$

or

$$q = -CV + K \tag{3.14}$$

where K is a constant of integration which can be nulled out, giving:

$$V = \frac{-q}{C} \tag{3.15}$$

Since q is proportional to the applied force, V is proportional to the pressure difference across the diaphragm.

In practice, the charge leaks away due to the non-infinite input impedance of the amplifier, giving responses similar to fig. 3.11c

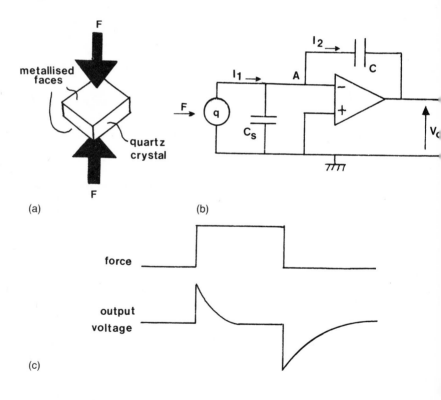

(a)   (b)   (c)

Fig. 3.11   Piezo-electric force transducer. (a) Piezo-electric effect. (b) Charge amplifier. (c) Low-frequency response of charge amplifier.

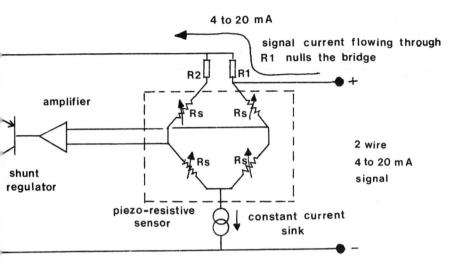

Fig. 3.12 Two-wire piezo-resistive transducer.

even with a FET amplifier. Piezo-electric transducers are therefore unsuitable for measuring *static* pressures. They do, however, have a fast response and are widely used for measuring fast *dynamic* pressure changes. A piezo-electric transducer, for example, can easily follow the pressure variations inside a car-engine cylinder head.

Suitably prepared crystals also exhibit a change in resistance with applied force: the piezo-resistive effect. Using this effect it is possible to construct a full Wheatstone bridge on a single chip, as fig. 3.12. Using a self-nulling balance circuit, an output current proportional to the applied force (and hence the differential pressure) is produced. Unlike the piezo-electric sensor, the piezo-resistive device can measure static pressure. A typical device is shown in fig. 3.13.

Both the piezo-electric and piezo-resistive devices are essentially *force* measuring sensors and need minimal displacement of the elastic sensing elements. This obviates the need for complex linkages and gives a more linear response as the output does not depend on the movement or elastic sensitivity of the mechanical sensing element.

Fig. 3.13 Piezo-resistive differential pressure transducer. Compare the mechanical simplicity with fig. 3.15.

3.5.   Force balancing systems

3.5.1.   Electrical systems

Friction and non-linear spring constants can cause errors in the elastic sensor/displacement transducers of section 3.3. The force balance principle provides a force to return the elastic sensor to a 'datum' position. The measurement of the force then indicates the pressure; providing a signal which is independent of the elastic characteristics of the sensor itself.

An electrical force balance transducer is shown in fig. 3.14. A differential pressure is applied to a conventional diaphragm, causing a movement of arm X which in turn moves arm Y. A displacement transducer (capacitive, inductive or LVDT) senses this movement. A servo amplifier applies a current to the solenoid at the other end of arm Y. The current is automatically adjusted until arm Y (and hence arm X and the diaphragm) are returned to the datum position.

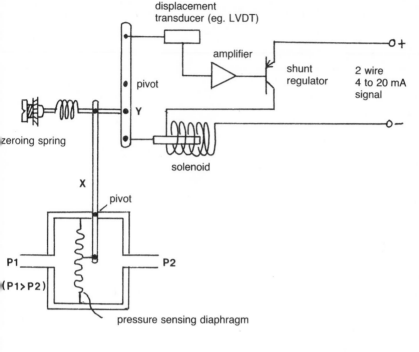

Fig. 3.14 Force balance pressure transducer.

The force exerted by the solenoid now balances the differential pressure force on the diaphragm. The solenoid force is directly proportional to the coil current, so an ammeter in the coil can be used to indicate differential pressure. A transducer using this principle is shown in fig. 3.15.

3.5.2. Pneumatic systems

A pneumatic force balance system is based on the ubiquitous flapper/nozzle (described in chapter 5, Volume 2) and is shown schematically in fig. 3.16. As before, the pressure differential initially moves the diaphragm, which in turn moves arms X and Y. As the flapper moves, say, towards the nozzle, the air loss through the nozzle decreases, causing the pressure in the bellows (and the output pressure) to increase.

The increased pressure on the bellows causes arm Y, and hence arm X and the diaphragm, to move back to the datum position. The output pressure is thus proportional to the pressure

Fig. 3.15   Force balance differential pressure transducer.

differential across the diaphragm. Like the electrical arrangement of section 3.5.1, the output pressure does not depend on the elastic constant of the diaphragm.

3.6. Pressure transducer specifications

The specification sheet for a pressure transducer will contain the usual details (hysteresis, error, linearity, etc.) plus other terminology that is unique to pressure measurement. A pressure transducer may be required to measure a differential pressure superimposed on a large pressure: flow measurement with an orifice plate in a high-pressure line, for example. The pressure that any transducer connection can stand with *respect to atmosphere* is called the *proof pressure* or the static pressure. A typical differential pressure transmitter may have a range of 0–0.1 bar and a proof pressure of 100 bar: 1000 times the range.

Overpressure is referenced to the actual measurement range, and is usually defined in three ways. First, and lowest, there is the overpressure which does not cause a zero or span shift. Next comes the overpressure which causes a zero or span shift. Finally, the overpressure which causes permanent damage, usually considerably higher. Devices can usually withstand a short pressure pulse, so often a dynamic overpressure is also specified, which is a high-pressure pulse (greater than the static permanent

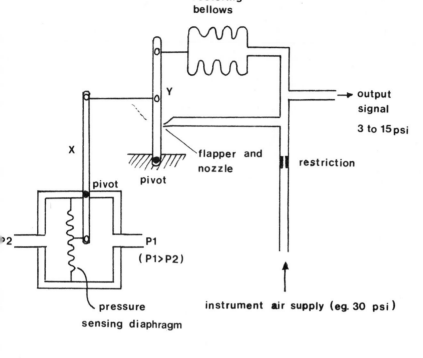

Fig. 3.16 Pneumatic force balance transducer.

damage pressure) specified for some maximum time. Figure 3.17 shows a common overload protection arrangement where a movable plug seals a bleed hole between HP and LP diaphragms.

Most pressure transducers employ elastic sensors, which inherently have a natural frequency and a tendency to give damped oscillations. The specifications for a device will give the natural frequency (typically 0.1–3 Hz) and the damping factor (usually adjustable). These factors determine the response to fast pressure changes (although the user should also remember the slugging effect of long pipe runs from plant to transducer).

3.7. Installation notes

If a long pipe run is used between plant and a pressure transducer, a significant volume of fluid will be moved in the pipes as the pressure changes. Time constants of several seconds can result, so piping runs should be kept short.

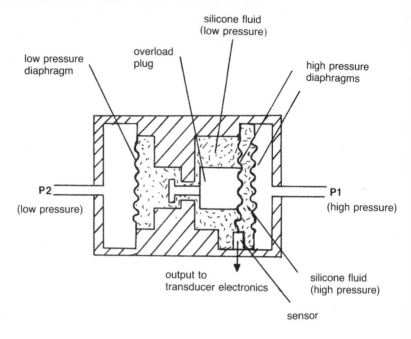

Fig. 3.17   Overload protection on differential pressure transducer.

Air or gas entrapped in piping can cause significant errors on liquid pressure measurement applications. Similarly, condensed moisture can form in piping on gas pressure systems. Ideally piping should rise from the pipe to the transducer on gas systems, and fall directly from the pipe to the transducer on liquid systems. Where this is not feasible, vent and drain cocks should be fitted, as in fig. 3.18. Under no circumstances should the piping form traps where air bubbles or liquid sumps can form (as in fig. 3.18c).

When a pressure transducer is situated below a pipe and liquid pressure measurement is being made, an error will be introduced from the static head of liquid in the signal pipe. The pressure error is $\varrho g h$, as shown in fig. 3.19. If this amounts to less than 10% of the pressure range being measured, it can usually be removed by zero adjustment of the transducer.

Pressure transducers will inevitably need to be changed on line, so isolation valves should be fitted. Figure 5.14 shows a typical arrangement for a differential transmitter. Where a low pressure range device is used with a high static pressure, care must be taken not to damage the device by overpressure caused by pipe

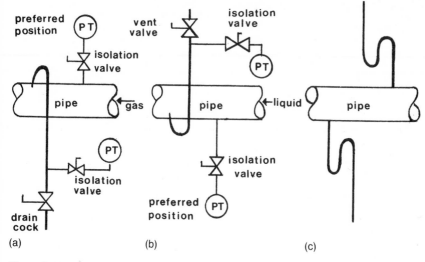

Fig. 3.18 Installation of pressure transmitters. (a) Gas pressure measurement. (b) Liquid pressure measurement. (c) Faulty piping with potential traps.

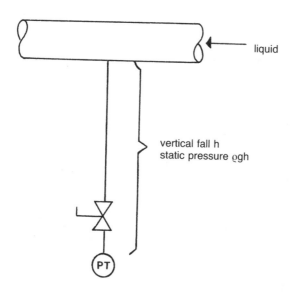

Fig. 3.19 Zero shift from static pressure.

pressure getting 'locked in' to one side. The bypass valve should always be opened first and shut last.

3.8. Vacuum measurement

3.8.1. Introduction

Vacuum measurement is normally made in terms of the height of a column of mercury supported by the vacuum (mmHg). Atmospheric pressure therefore corresponds to about 760 mmHg, and 'absolute' vacuum to 0 mmHg. The term 'torr' is generally given to 1 mmHg.

Conventional absolute pressure transducers are usable without difficulty down to 20 torr, and to 1 torr with special diaphragms. At lower pressures, the scales should be considered logarithmic, i.e. the range 1–0.1 torr equates in range to 0.1–0.01 torr. Measurement down to below 10^{-7} torr is sometimes required.

3.8.2. Pirani gauge

Heat is lost from a hot body by conduction, convection and radiation. If a hot wire is surrounded by a gas, the first two losses are pressure dependent because the heat transfer depends on the number of gas molecules per unit volume.

In the Pirani gauge, constant energy is supplied to a wire in the vacuum to be measured. As the level of vacuum varies, the differing heat losses cause the temperature of the wire to change. In fig. 3.20a, the temperature of the wire is sensed by a

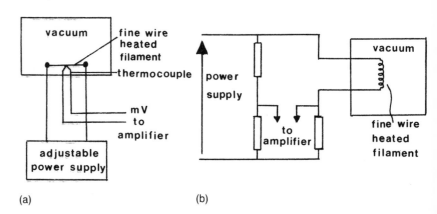

(a)   (b)

Fig. 3.20   The Pirani vacuum gauge. (a) Thermocouple sensor. (b) Combined filament and resistance sensor.

thermocouple; the voltage indicating the pressure. Alternatively the temperature change can be sensed by the change in resistance of the wire with temperature, as in fig. 3.20b, or by using a self-balancing bridge.

The range of the Pirani gauge can easily be changed by altering the energy supplied to the wire. A typical instrument would cover the range $5{-}10^{-5}$ torr in three ranges. Care must be taken not to overheat the wire, which causes a change in characteristics due to surface oxidation or even wire breakage.

3.8.3. Ionisation gauge

The ionisation gauge (shown in fig. 3.21) is superficially similar to an electronic triode, Electrons emitted from the heated filament cause ionisation current to flow which is proportional (within limits) to the absolute pressure. The current can be measured with high-sensitivity microammeters. Ionisation gauges cover the range $10^{-3}{-}10^{-12}$ torr.

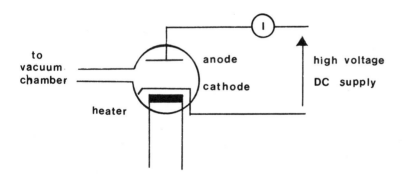

Fig. 3.21   The ionisation vacuum gauge.

Chapter 4
Position transducers

4.1. Introduction

The measurement of position is of fundamental importance in many systems: steel rolling mills, radio astronomy and numerically controlled machine tools all require accurate control of the position of some equipment. In addition, transducers for pressure, weight, temperature, level and other physical variables often convert the measurement to a displacement which is then converted to an electrical signal by a position or displacement sensor.

Position transducers can measure linear displacement or angular displacement, although the two can be easily interchanged by screws, rack-and-pinions, and similar mechanisms. The linear position of the tool on a lathe, for example, can be measured by an angular position transducer connected to the lead screw.

The measurement of position can be *absolute* or *incremental*. An absolute transducer measures the position at all times with respect to some fixed datum. An incremental transducer gives a signal corresponding to distance moved, and as such does not correctly indicate position after a power fail. The simplest, and commonest, incremental encoder is a pulse counter, one example of which is shown in fig. 4.1. In this application horizontal movement of a hydraulic ram is measured by counting pulses off a toothed wheel via a photocell, the wheel being turned by the oil passing to and from the ram. Incremental measurement is simple and cheap, but a datum must be established after each power-up and at regular intervals to avoid ambiguity. Care must also be taken to prevent cumulation of errors each time a reversal of direction occurs.

Fig. 4.1 Incremental position measurement by toothed wheel & photocell.

4.2. The potentiometer

The simplest displacement transducer is the humble potentiometer. The wiper of the potentiometer is mechanically linked to the object whose displacement is to be measured, and an electrical output voltage directly proportional to wiper position is produced. Potentiometers, or 'pots' as they are more commonly known, can directly measure angular or linear displacement as shown in fig. 4.2.

The circuit of a pot-based displacement transducer is shown in fig. 4.3. The ends of the potentiometer are connected to a stable voltage, V_i. If L is the maximum displacement of the pot, and d the slider position, the unloaded output voltage will be:

$$V_o = \frac{d}{L} \cdot V_i \qquad (4.1)$$

It is often convenient to consider the fractional displacement x, given by

$$x = \frac{d}{L} \qquad (4.2)$$

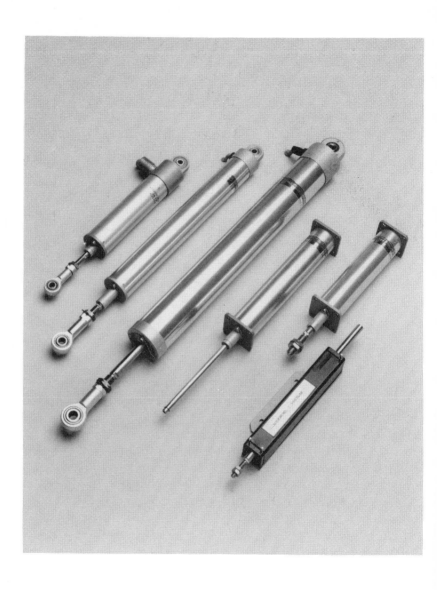

Fig. 4.2 Industrial potentiometers. (a) Linear – see opposite page. (b) Rotary – see above. Note universal joints at ends of linear potentiometers. (Photos courtesy of Penny & Giles.)

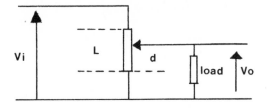

Fig. 4.3   Simple potentiometer circuit.

where $0<x<1$. Equation 4.1 can be rewritten:

$$V_o = x.V_i \tag{4.3}$$

The above equations assume no current is drawn from the slider. If a finite load is connected, the relationship between position and output voltage becomes non-linear, as shown in fig. 4.4a. For a load resistor R_L and potentiometer resistance R, the output voltage for a fractional displacement x is:

$$V_o = \frac{xV_i}{1 + x(1-x)(R/R_L)} \tag{4.4}$$

The maximum error occurs at $x = 2/3$, and if R_L is significantly larger than R, the error is about $25R/R_L\%$, as a percentage of true value or $15R/R_L\%$ as a percentage of full scale. A 10 k-ohm pot with a 100 k-ohm load, for example, will have an error of 1.5% FSD due to loading.

Loading errors are reduced by having a high impedance load or a low-value potentiometer, but the latter is limited by the maximum allowable dissipation in the potentiometer itself. The dissipated power is given by:

$$P = V_i^2/R \tag{4.5}$$

A typical value of maximum power for a potentiometer is 1 W, so for a 10-V supply the minimum potentiometer value would be 100 ohm. The maximum dissipated power is usually quoted for some temperature (often 50 °C) and the potentiometer has to be derated considerably for higher ambient temperatures.

Loading errors are further augmented by linearity and resolution errors from the potentiometer itself, and system errors due to backlash and friction. Pots inherently have finite resolution; from the wire size for wire-wound pots and the grain size for metal film and cermet pots. The common carbon

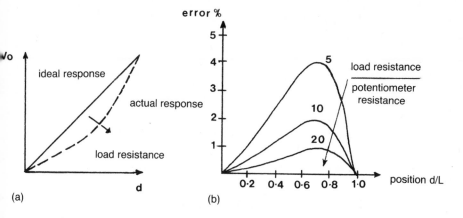

Fig. 4.4 Effect of non–infinite load resistance on output signal. (a) Position/output voltage relationship. (b) Error induced for various values of load.

potentiometer (used for adjustment purposes) tends towards a dirty track with much movement and should be avoided. With care (and some expenditure) resolutions and linearities of 0.01% can be obtained.

There are safety implications in using potentiometers for position measurement in closed loop systems. Examination of fig. 4.3 will show that a dirt spot on the track giving an open circuit wiper, or a track break above the wiper will cause the output to read full-scale low, whereas a track break below the wiper will cause the output to read full-scale high. The symptom of a track break or dirty track is often a high-speed dither of a closed loop system about the failure point of the pot.

The pot is essentially a mechanical device and suffers from friction effects. As a result it puts loading on the measuring device which induces errors and increases effects such as backlash. Careful mechanical design of linkages and gearboxes is needed to reduce mechanical errors to an acceptable level.

Being a mechanical moving device, the life of a pot is limited, failure usually being caused by wear of the wiper or the track. The life is reduced further if the potentiometer operates permanently over a small range of its travel.

Potentiometers with a specified non-linear response can be manufactured by varying the track resistance along its length. Logarithms, square root and trigonometrical (sine, cosine) responses are readily available.

4.3.   Synchros and resolvers

4.3.1.   Basic theory

Figure 4.5a represents a simple transformer, with an AC input voltage V_i. The output voltage V_o will be KV_i where K is a constant dependent on the turns ratio and the losses in the transformer. In fig. 4.5b the secondary has been rotated through 90°. The flux from the primary induces equal and opposite voltages in the secondary, giving a net output voltage of zero. In fig. 4.5c the secondary has been rotated further to 180° from the initial position. The output voltage will be the same as fig. 4.5a, but the output voltage is *antiphase* to the input, i.e. $V_o = -KV_i$.

Figures 4.5a–c are special cases of fig. 4.5d where the secondary is rotated to an angle θ with respect to the primary. The output voltage will be given by:

$$V_o = KV_i \cos \theta \qquad (4.6)$$

This relationship is shown in fig. 4.6. It is important to appreciate the significance of equation 4.6. The output voltage can only be in phase with the input (cos θ positive; $\theta > 270°$ or $\theta < 90°$) or

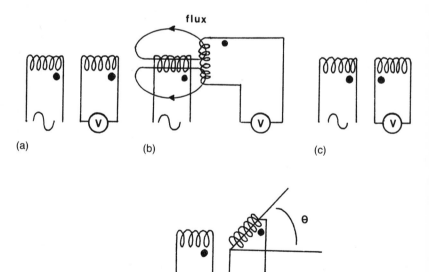

Fig. 4.5 Basic theory of resolvers and synchros. (a) Coils aligned. (b) Coils at 90°. (c) Coils aligned but reversed. (d) Generalised case.

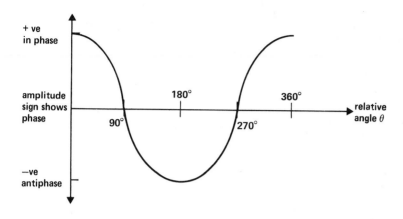

Fig. 4.6 Relationships between relative angle and output voltage, amplitude and phase.

antiphase with the input (cos θ negative; 90°<θ<270°). At the points θ = 90° and θ = 270°, the output voltage is theoretically zero (although in practice leakage flux and manufacturing tolerances will cause a very small voltage to be present). Output waveforms for various angles are shown in fig. 4.7.

Figures 4.5–7 form the basis of two types of position transducer, synchros and resolvers. These are both angular measuring devices, but can be used to measure linear displacement by suitable mechanical linkages.

4.3.2. Synchro torque transmitter and receiver

The synchro's transmitter and receiver pair (often described by the trade name 'Selsyn' link) is widely used where remote indication of some angular reading is required. Remote indication of a mechanical weigh scale as shown in fig. 4.8 is a typical application. The synchro transmitter receiver replaces a mechanical linkage as there is a 1 : 1 relationship between the transmitter input shaft and the receiver output shaft. The devices are called *torque* units because the receiver exerts a positioning torque directly on its shaft. *Control* units, described in the next section, are used as part of a position control system.

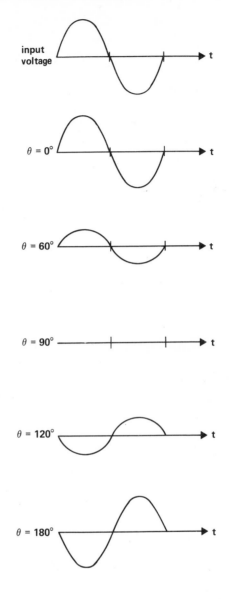

Fig. 4.7 Output voltage/time relationship for various angles.

A torque transmitter, shown in fig. 4.9a, looks superficially like a small electric motor. Internally it has two major components, a stator with three separate windings at 120° star connected, and a rotor which can be considered to be one winding. The rotor connections are bought out via slip rings. The transmitter can

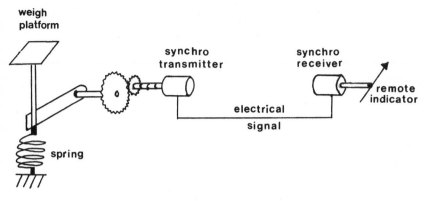

Fig. 4.8 Synchro transmitter/receiver.

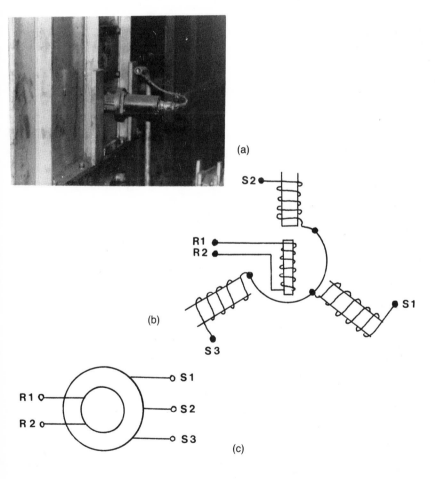

Fig. 4.9 The synchro torque transmitter. (a) Typical resolver/synchro.
(b) Construction of a torque transmitter. (c) Schematic representation.

therefore be represented as fig. 4.9b. The schematic of fig. 4.9c is sometimes used.

In use, an AC supply (usually 400 Hz or 50 Hz, 110 V) is applied to the rotor connections R_1, R_2. Voltages will be induced in the stator coils whose phase and magnitude uniquely define the angle of the rotor. To use these voltages a torque receiver is connected to the transmitter.

The construction of a torque receiver is almost identical to that of a torque transmitter. Electrically they are identical but the torque receiver has low friction bearings and the addition of a damping disc to the rotor to prevent oscillations and limit acceleration. The receiver terminals are connected to the corresponding transmitter terminals as shown in fig. 4.10.

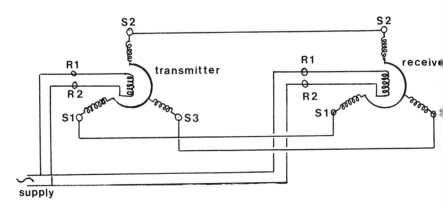

Fig. 4.10   A synchro transmitter–receiver link.

The AC voltage applied to the transmitter rotor induces voltages in the stator windings as explained above, which causes current to flow through the stator windings of the receiver. These currents produce a magnetic field at the receiver which is aligned at the same angle as the transmitter rotor.

The AC voltage applied to the receiver rotor also produces a magnetic field. If this field does not correspond in direction with the stator-produced field, a torque (caused by magnetic repulsion/attraction) will be experienced by the receiver rotor. This torque will cause the rotor to rotate until the rotor and stator fields are aligned; i.e. the receiver rotor is at the same angle as the transmitter rotor.

If the transmitter rotor is rotated to some new angle, the

magnetic field from the receiver stator will also move causing the receiver rotor to rotate until it is again at the same angle as the transmitter rotor. The receiver shaft thus follows the movements of the transmitter shaft.

If the receiver is purely driving an indicator (which requires little or no torque), positioning accuracy is about 0.25°. If the receiver is required to produce torque (to move a pilot valve, say), the torque is obtained by a standing error between the receiver rotor position and the receiver stator field. If more than a nominal torque is required, a closed loop system using a control transformer is needed, as described in section 4.3.4.

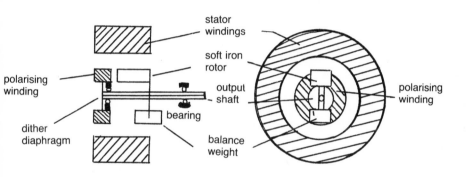

Fig. 4.11   The Magslip indicator.

A more accurate indicator, called a Magslip, is shown in fig. 4.11. The indicator is connected to a balanced L-shaped soft iron rotor. The stator is constructed of three windings identical to a synchro receiver. A fixed polarising coil induces the field into the rotor, which aligns itself with the stator field. The polarising coil also exerts a pulsating pull on the rotor, which causes the rotor to 'dither' slightly on its bearing. This, along with the absence of slip rings, reduces the effect of stiction and friction and gives increased accuracy.

4.3.3.   Zeroing of synchros and lead transpositions

The zero position of a synchro is defined as the rotor R_1 end being aligned with stator winding 2 as shown in fig. 4.10. In this position a minimum voltage exists between S_1 and S_3 (although note that another, identical, minimum occurs at 180° from zero). In practice, however, actual knowledge of synchro zero is not required: all that is needed is to align the receiver shaft with the transmitter shaft. This can be achieved by rotating the whole body of the transmitter or receiver.

Synchros are manufactured with a small lip, as shown in fig. 4.12, and are mounted into an accurately machined hole. Small mounting brackets locate on to the lip and allow the synchro body to be turned by hand for zeroing. For more accurate zeroing the synchro can be mounted in a frame which can be rotated by a gear or screw-thread mechanism. Electrical zeroing can also be performed by the differential transmitter, described in section 4.3.5.

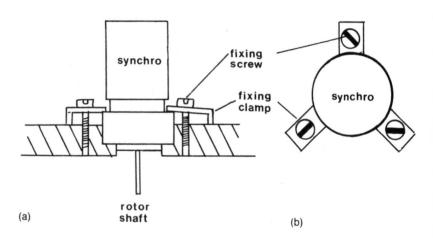

Fig. 4.12 Mounting and zero adjustment of synchros. (a) Side view. (b) Top view.

The zero and direction of receiver shaft rotation can be altered (deliberately or accidentally) by interchanging of leads. The results of a few of the many possible interchanges are shown in fig. 4.13. The most important of these is the interchanging of the rotor leads on one device (which shifts the zero by 180°) and the interchanging of two stator leads (which reverses the direction of rotation of the receiver). Where reversed direction is required, the preferred exchange is S_1–S_3 which leaves the nominal synchro zero unchanged.

The power factor of a synchro link is poor because of the

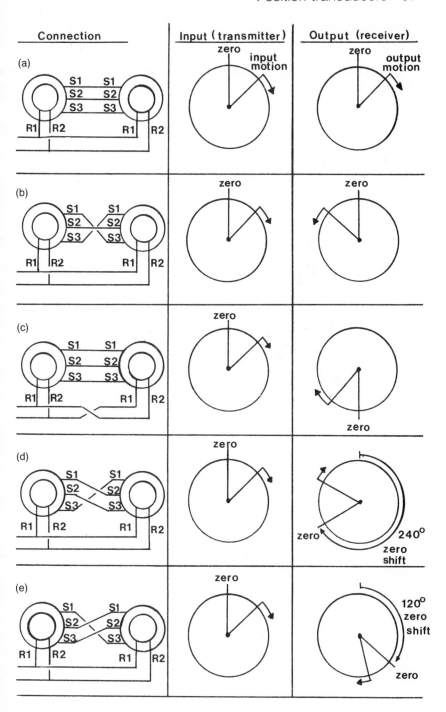

Fig. 4.13 Various synchro connections. (a) Standard. (b) Crossed stator. (c) Crossed rotor. (d) Cyclic shift. (e) Cyclic shift.

magnetising current of the stator winding, which can be in excess of 60% of the stator current. This leads to a loss of accuracy, but can be overcome by the addition of three suitable matched capacitors across the stator lines.

4.3.4. Control transformers and closed loop control

A torque receiver only produces a small torque, and can only drive small loads such as indicators or pilot valves. To drive larger loads such as gun turrets or rolls in a steel mill, a closed loop system is required, with a synchro-based error measuring device to indicate the error between the set point and the actual position.

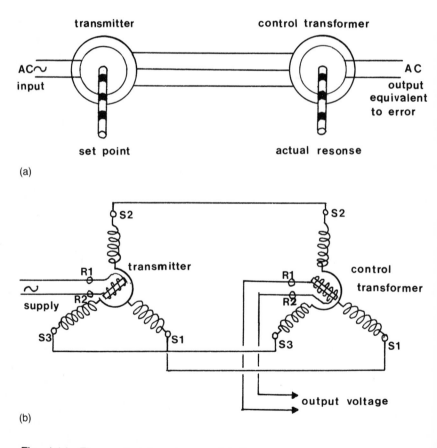

Fig. 4.14 The control transformer. (a) Control transformer operation. (b) Control transformer connection.

The basis of such a closed loop system is the *control transformer*, which is used as shown in fig. 4.14a. The set point is set mechanically on the shaft of a transmitter, which gives an electrical set point on S_1–S_3. The actual position is set on the shaft of the control transformer; the error is an AC signal of the same frequency as the driving supply, with the magnitude indicating the size of the error, and the phase (in phase or antiphase) indicating the direction.

Internally, a control transformer is similar to a transmitter, so fig. 4.14a can be redrawn as fig. 4.14b. As explained in the previous section, at any transmitter rotor angle, a magnetic field will be produced at the same angle by the stator windings in the control transformer. This field will indicate a voltage in the rotor windings which will be a minimum when the rotor windings are at right angles to the field. The aligned position for the control transformer is therefore 90° displaced from the aligned position for a receiver. (There will, of course, be a second aligned position 180° displaced. Care must be taken to avoid ambiguity.)

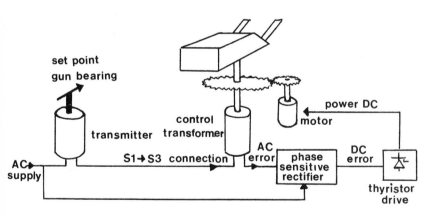

Fig. 4.15 Position control based on control transformer.

The basis of a closed loop system using a control transformer is shown in fig. 4.15. The set-point angle is set on the transmitter, and the AC error signal between the actual and set point is given by the control transformer. The AC error signal is converted to a DC error signal by a phase-sensitive rectifier (giving a positive signal for in-phase input, negative signal for antiphase). The DC error signal becomes the speed reference for a thyristor drive, controlling the speed of the positioning motor.

There are various phase-sensitive rectifier circuits, and the commonest are shown in fig. 4.16. The circuit of fig. 4.16a, known as a Cowan circuit, is a half-wave rectifier. When V_{ref} is negative, the diodes are back biased and V_o follows V_i from the control transformer. When V_{ref} is positive, the diodes conduct, shorting out the signal. The output therefore consists of positive half cycles if the input is in phase with V_{ref}, and negative half cycles when it is antiphase. Resistors R_1 and R_2 limit the currents through the diodes when they are conducting.

The Op Amp circuit of fig. 4.16b gives a full wave rectified output. This is a modification of the inverting/non-inverting amplifier (described in chapter 1, Volume 2), with the FET switch being operated on alternate cycles by V_{ref}. If V_i is in phase

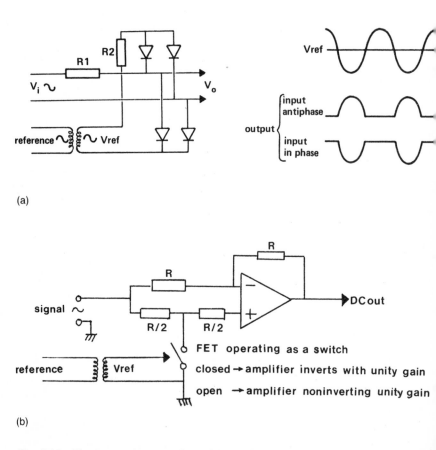

(a)

(b)

Fig. 4.16   Phase-sensitive rectifiers. (a) The Cowan half-wave phase-sensitive rectifier. (b) Full-wave rectifier based on operational amplifier.

with V_{ref}, a positive full wave output is given; if it is antiphase a negative output.

Both of the circuits shown in figs. 4.16a and b need filtering to give a DC signal. This filtering will degrade the frequency response of the position control system, and should be designed for as high a cut-off frequency as possible. For this reason, 400 Hz AC supplies are used in preference to 50 or 60 Hz supplies when a fast response is required (400 Hz synchros are also smaller than their 50 Hz counterparts).

Figures 4.14 and 4.15 imply that a control transformer is used with the same transmitter as a torque receiver. In practice, low stator currents are required for maximum accuracy. Control transformers are therefore wound with fine wire to give a higher impedance and are used with *control transmitters* which are similar to torque transmitters but again wound with high impedance wire.

The error in a closed loop synchro system can be reduced by gearing up at both ends of the link, as shown in fig. 4.17a. This reduces the error in proportion to the gear ratio; but introduces the probability of false positions (with an 18 : 1 ratio, for example, the control transformer could home every 20° of the output position).

To overcome this ambiguity while maintaining the accuracy of the geared system, a coarse/fine synchro system as in fig. 4.17b is commonly used. Assuming an 18 : 1 gear ratio, the system drives on the coarse 1 : 1 chain until the changeover circuit detects a coarse error of less than 20°. At this point it changes over and drives home with the increased accuracy of the fine chain. As the final positioning is done on the geared fine chain, increased accuracy is obtained without the possibility of false home positions.

4.3.5. Differential synchros

Differential synchros are constructed with three rotor windings, as in fig. 4.18. There are two types of differential synchro: differential transmitters and differential receivers. Both are electrically similar, but the differential receiver incorporates flywheel damping (as outlined earlier for torque receivers).

The differential transmitter is connected between a transmitter and a receiver (or control transformer), as in fig. 4.19a. The

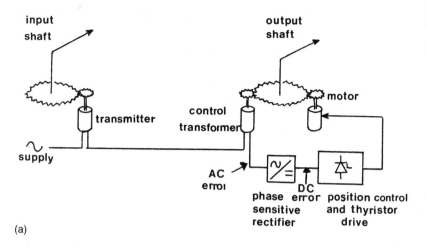

(a)

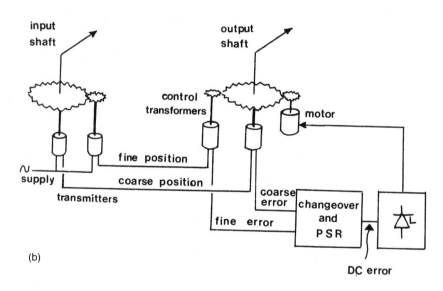

(b)

Fig. 4.17 Methods of increasing accuracy on synchro links. (a) Increasing accuracy with gearing. (b) Coarse/fine changeover system.

signal it transmits is the difference between the signal on its stator lines and its own rotor. On fig. 4.19a the stator lines are indicating 90° and the rotor is at 30°, so the new transmitted signal is 90° − 30° = 60°. Differential transmitters are often used for zeroing purposes in place of the less elegant zeroing of fig. 4.12. By interchanging S_1, S_3 and R_1, R_3, the differential transmitter adds angles as in fig. 4.19b.

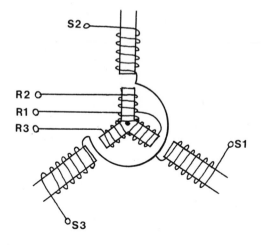

Fig. 4.18 The differential synchro.

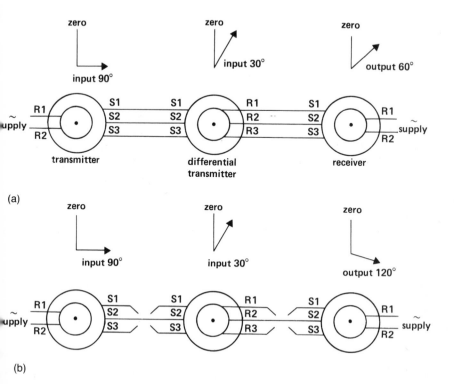

(a)

(b)

Fig. 4.19 Differential transmitter applications. (a) Differential transmitter subtracting angles. (b) Differential transmitter adding angles.

The differential receiver takes two electrical signals as shown in fig. 4.20, and gives a rotor displacement equal to the difference between the two electrical signals. It is not as widely used as the preceding synchro devices.

Note the difference between the differential transmitter and receiver. The transmitter has an electrical and an angular input and an electrical output. The receiver has two electrical inputs and an angular output.

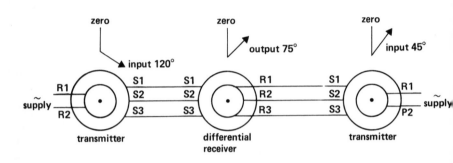

Fig. 4.20   The differential receiver.

4.3.6.   Synchro identification

There are five common sizes of synchro: 23, 18, 15, 11, 09. This figure denotes the outside diameter in tenths of an inch. There are two standard codes for synchros, MIL spec and Muirhead.

A MIL spec code has the form 11CT4b. The first numbers specify the size, followed by a code indicating the type as below:

CX	control transmitter
CT	control transformer
CDX	control differential transmitter
TX	torque transmitter
TR	torque receiver
TDX	torque differential transmitter
TDR	torque differential receiver

The next digit indicates the frequency, 4 for 400 Hz, 6 for 60 Hz. The final letter is a modification code.

The Muirhead code has the form 18M2B3. The first two digits

specify the size, as before, followed by a letter indicating the accuracy:

M	Military (standard)
C	Commercial (reduced accuracy)
R	Reference (high accuracy)

The next digit specifies the synchro type:

1	control transmitter
2	control transformer
3	control differential transmitter
4	torque receiver
5	torque differential transmitter
8	torque differential receiver
9	torque transmitter

The final letters/digits are a modification code.

4.3.7. Resolvers

Resolvers (coded R in the MIL code) have two stator coils at right angles and a rotatable rotor coil, as in fig. 4.21a (often two rotor coils at right angles are provided, but only one is used for position control and indication). The voltages induced into the two stator coils are simply:

$$E_1 = K E_{in} \cos \theta$$
$$E_2 = K E_{in} \sin \theta$$

$$(4.7)$$

Resolvers are used for coordinate conversion and conversion from rectangular to polar coordinates. They are also widely used with solid-state digital converters. One advantage of resolvers is that $E_1^2 + E_2^2$ is a constant, which makes it easy to check for an open circuit signal.

Resolvers can also be used with AC applied to the stator windings and the position read off the rotor winding. In this operating mode the stators are fed with AC signals with a 90° phase shift, as in fig. 4.21b. If the frequency of the drive signals is $\omega/2\pi$, they can be represented as $\sin(\omega t)$ and $\cos(\omega t)$. The output rotor signal is then:

$$V_o = K V_i(\sin \omega t \cos \theta + \cos \omega t \sin \theta) \qquad (4.8)$$

where V_i is the input signal amplitude (both stators assumed the same), θ is the rotor angular position, and K is a coupling

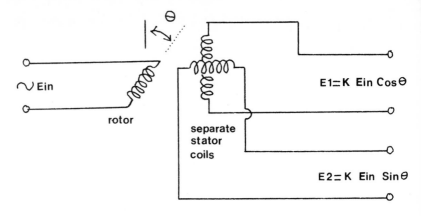

(a)

(b)

Fig. 4.21 The resolver. (a) Drive signal applied to rotor, position read from stators. (b) Drive signal applied to stators, position read from rotor.

constant.

By trigonometrical manipulation:

$$V_o = K V_i \sin(\omega t + \theta) \qquad (4.9)$$

which is a constant-amplitude signal with frequency identical to the drive signal but a phase shift equal to the rotor position. This can be converted to a DC signal by a phase-sensitive rectifier.

A close relative of the resolver is the inductive potentiometer of fig. 4.22. This gives an output of $K \cos \theta$ or, with special construction, $K \theta$. The output signal is only unique over the range $0 < \theta < 180°$, so the inductive potentiometer has a restricted angular range.

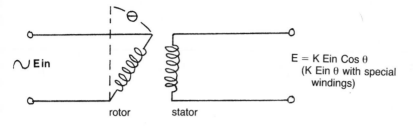

$$E = K \, Ein \, Cos \, \theta$$
(K Ein θ with special windings)

rotor   stator

Fig. 4.22   The inductive potentiometer.

4.3.8.   *Solid-state converters*

Hybrid encapsulated circuits are available to convert the stator signals from synchros or resolvers direct to a digital number in binary, BCD, or angle in BCD form. A typical device, shown in fig. 4.23, can resolve to one part in 4096 (12 binary bits). Coarse/fine changeover circuits are also available.

These devices, known as SDCs (for synchro-to-digital converter), are particularly attractive for computer-based position measurement/control systems. Resolver SDCs usually drive as in fig. 4.21b and incorporate the quadrature oscillator.

4.4.   Shaft encoders

4.4.1.   *Absolute encoders*

An absolute-position shaft encoder can be represented by fig. 4.24a. An angular displacement is applied to the shaft input, and a set of parallel digital output lines give a unique and unambiguous indication of the shaft position. The digital outputs can be binary (as in fig. 4.24b), BCD or angular-coded BCD (0–360, say, for one shaft revolution). Resolution of one part in 4000 (12-bit binary) is easily obtainable.

Most absolute shaft encoders use optical methods similar to fig. 4.25. For simplicity of explanation, fig. 4.25 has only a resolution of one part in 16, but the techniques are identical for higher resolutions. A transparent disc (fig. 4.25a) has coded tracks according to some required pattern. In fig. 4.25 a binary pattern is used, but in practice a unit distance code, described below, is usually employed. The disc is illuminated on one side, and

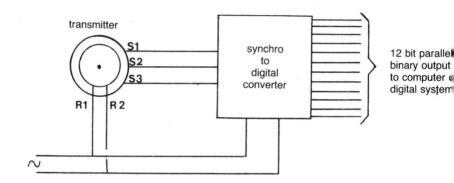

(a)

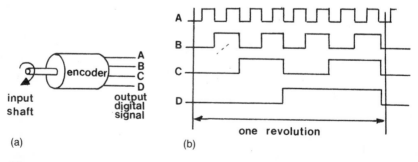

(b)

Fig. 4.23 Solid-state synchro converters. (a) SDC circuit arrangement. (b) Photo of separate SDC module. Analog Devices 12-bit SDC/RDC 1767 and 14-bit SDC/RDC 1768 are synchro- and resolver-to-digital converters. The hybrid devices combine high tracking rates with a velocity output and internal transformer isolation. (Photo courtesy of Analog Devices.)

Fig. 4.24 The shaft encoder. (a) Schematic. (b) 4-bit output signal.

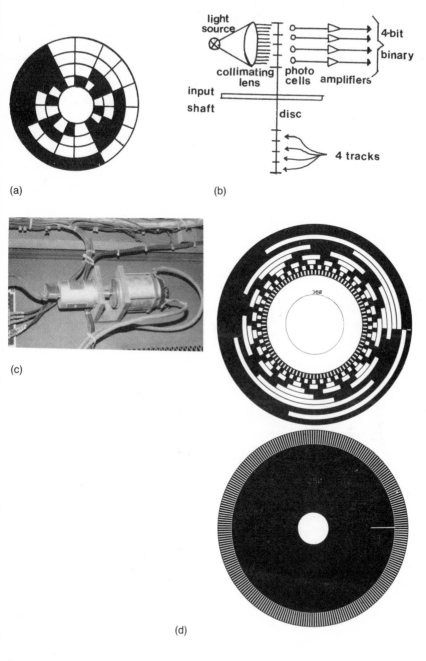

Fig. 4.25 Construction of shaft encoders. (a) 4-Bit coded disc (inner track, Least Significant Bit). (b) Side view. (c) Two linked position transducers, a Selsyn driving a shaft encoder. (d) Commercial absolute and incremental encoder discs. (Photo courtesy of Hohner Automation.)

photocells sense the track pattern on the other side. The track coding corresponding to the current angular position can be read directly off the photocell amplifiers. Figure 4.25a shows a four-track disc. Commercial units employ up to twelve tracks; a typical example is shown in fig. 4.25d. An additional unmasked photocell is often included to act as a lamp failure detector and to give a reference level for the track photocells.

A simple binary-coded shaft encoder can give false readings as the outputs change state. Suppose a four-track encoder is going from 0111 to 1000. It is unlikely that all PECs will change together, so the outputs could go 0111→0000→1000 or 0111→1111→1000, or any other combination of four bits.

There are two solutions to this problem. The first is commonly used where a BCD output is required, and utilises an additional photocell to inhibit changes in the output around transition points. The additional track, often called an anti-ambiguity track, is arranged as shown in fig. 4.26a. The output from the corresponding photocell is connected to hold/follow latches, as in fig. 4.26b. Usually all the circuit is contained within the encoder body.

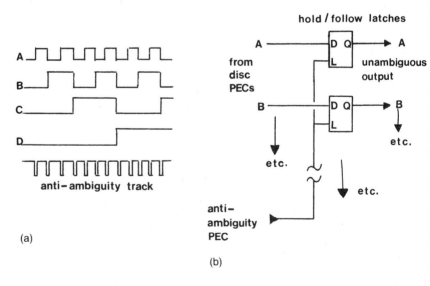

Fig. 4.26 Use of anti-ambiguity track. (a) Disc coding. (b) Decoding circuit.

Where a binary output is needed, a Gray code disc is used. In this coding, only one bit changes at a time so there is no ambiguity. Because only one bit changes, the term *unit distance code* is often used. A four-bit Gray code, for example, goes:

Decimal	Gray code
0	0000
1	0001
2	0011
3	0010
4	0110
5	0111
6	0101
7	0100
8	1100
9	1101
10	1111
11	1110
12	1010
13	1011
14	1001
15	1000

It will be noted that the code is symmetrical about the 7/8 transition, and for this reason such codes are often called reflected codes. Construction of unit distance codes and conversions to and from binary are described in chapter 3 in Volume 2.

Most shaft encoders today use optical techniques, but other designs are available. A cheaper but less robust encoder can be made with a copper-covered disc etched to the required pattern. The track data is then read off by small copper or carbon brushes. These brush shaft encoders have no internal circuits and can work with a wide range of signal levels. The inevitable brush friction, however, gives a relatively high shaft torque and a limited life due to brush or track wear.

4.4.2. Incremental encoders

Incremental encoders are the simplest position-measuring devices but do not give an unambiguous indication of position as do potentiometers, synchros or absolute shaft encoders. Incremental encoders give, instead, a pulse output with each pulse corresponding to some predetermined distance. These pulses are counted by some external counter which then indicates the distance moved.

In many applications it is possible to construct incremental encoders as part of the plant. Figure 4.1, for example, uses a toothed wheel and a photocell to measure the extension of a hydraulic ram, the wheel being turned by fluid going to the ram through a hydraulic motor. In fig. 4.27 a proximity detector is used to count revolutions of the drive shaft to a screw jack to measure the jack height.

Fig. 4.27 Incremental encoder based on a proximity detector & toothed wheel.

Incremental encoders, though cheap and simple, do have some disadvantages. The first is that position measurement is normally lost after a power fail (unlike previous devices which correctly, and unambiguously, give the position on resumption of power). Systems using incremental encoders must therefore incorporate some way of establishing a datum position from which counting can recommence.

Simple devices, such as fig. 4.1 and fig. 4.27 can also gain or lose a count on each reversal of direction if the counter sense is solely selected by the *supposed* direction of movement. Such devices are particularly prone to cumulation of errors during overshoots and oscillations. More complex encoders, described below, give direct indication of count direction.

Commercial encoders are based on the principle of fig. 4.1 with its light source, toothed wheel and single photocell. One major problem, though, is that for fine resolution the teeth, and hence the photocell head, become very small. The toothed wheel, therefore, is usually replaced by a Moiré fringe assembly which is best understood by reference to fig. 4.28. Moiré fringes give excellent resolution without the need for very small photocells. A linear displacement device is shown first for ease of explanation.

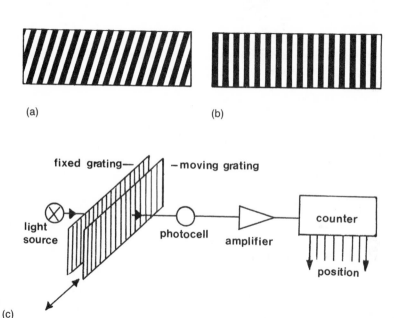

Fig. 4.28 Linear Moiré fringe position transducer. (a) Fixed grating. (b) Moving grating. (c) Arrangement.

The unit consists of two transparent plates which have grating patterns as shown in figs. 4.28a and b. These are arranged as in fig. 4.28c with a lamp and photocell. As the moving grating moves with respect to the fixed grating, diagonal dark/light areas move *vertically* past the photocell giving the required pulse output train. The light patterns are known as Moiré fringes, and the pulse/distance relationship depends on the number of gratings on the fixed and moving plates. Devices similar to fig. 4.28c are widely used on numerically controlled machine tools, giving a resolution of 0.01 mm with a total travel of several metres. Where a long travel is needed, a reflective grating is often attached to the body of the machine, and the photocell moves with the carriage.

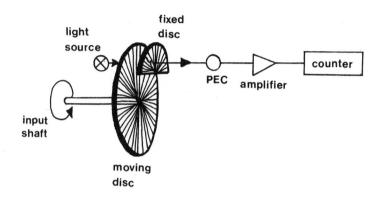

Fig. 4.29 Angular incremental encoder. Note that only a small segment of the fixed disc is needed.

There are several ways of producing Moiré fringes with angular rotation, but all use a rotating disc moving in front of a fixed disc, as in fig. 4.29. The angular equivalent of fig. 4.28 uses one disc with skewed lines. A technique called a vernier Moiré fringe device uses N segments on one disc and $N + 1$ on the other. As the moving disc rotates, m light/dark transitions are produced where m is the number of segments on the moving disk. Other techniques use two identical discs, either on the same axis or with the fixed disc slightly offset. The aim or all of these is to produce a known number of light/dark transitions per revolution.

With both the linear device of fig. 4.28 and the angular devices of fig. 4.29, the actual photocell output signal will not be a square wave, but will probably be somewhat sinusoidal with slow rising and falling edges. The photocell signal is therefore usually passed through a Schmitt trigger circuit to give a clean pulse output with sharp edges.

A single pulse output train carries no information as to direction. In simple low-accuracy applications, the direction can be obtained indirectly: the counter driven by fig. 4.27, for example, could have its count direction selected by the state of the reversing contactor feeding the AC motor driving the screw. There is still the possibility of a missed or gained pulse, however, so regular zeroing at some datum position is needed.

An incremental incoder can provide directional information with the addition of a second photocell arranged so its output is shifted by 90° in position, as in fig. 4.30. Let us call these two signals A and B. For clockwise rotation, let us say output A leads B by 90°, as in fig. 4.30a. If the direction is reversed, output B will lead A by 90°. Clockwise rotation is indicated by '(positive edge on B) and A high'. Similarly, anticlockwise rotation is indicated by '(positive edge on A) and B high'. These signals can be generated by RC differentiation and used to set or reset a flipflop, as in fig. 4.31. The output of this flipflop becomes the direction signal for the counter indicating position (digital circuits are described in chapter 3, Volume 2). The counter will follow reversals in direction without cumulative error (but zeroing at a datum position will still be required after a power failure). Obviously a similar technique can be applied to the linear incremental encoder of fig. 4.28.

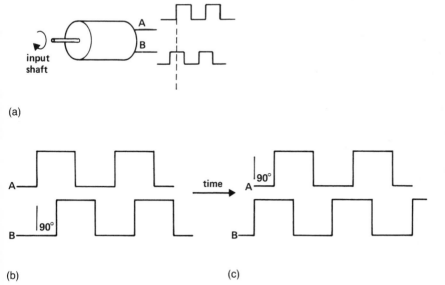

(a)

(b)   (c)

Fig. 4.30 Incremental encoder with two outputs for directional information. (a) Directional incremental encoder. (b) Clockwise rotation. (c) Anticlockwise rotation.

ICH1-5

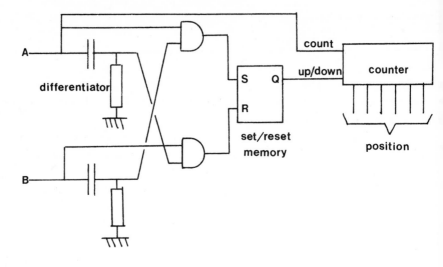

Fig. 4.31 Using a two-output incremental encoder.

4.5. Small displacement transducers

4.5.1. Introduction

Previous transducers can be used to measure distances and displacements of any required magnitude by suitable mechanical coupling. This section deals with transducers that measure small displacements. These can be position transducers in their own right (measuring the position of a pilot valve spool, for example) or as an integral part of a transducer measuring some other process variable. The commonest examples of the latter are pressure transducers which invariably convert the pressure to movement of a diaphragm.

4.5.2. Linear variable differential transformer (LVDT)

The commonest small displacement transducer is the LVDT, shown schematically in fig. 4.32a. The unit consists of a transformer with two secondary windings and a movable core. With the core in the centre position, the voltage in the two secondaries will be equal, and the output voltage will be zero.

If the core moves in a positive direction, voltage V_1 will

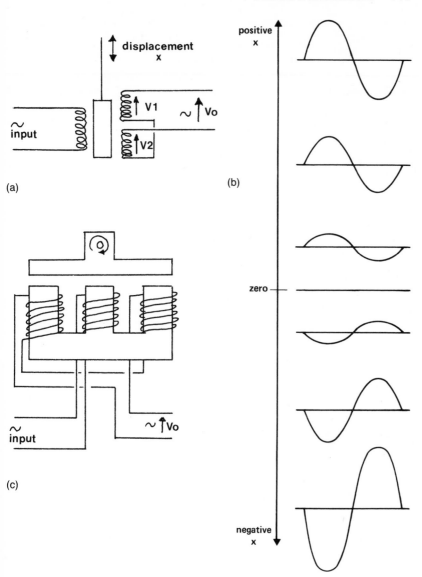

Fig. 4.32 The linear variable differential transformer (LVDT). (a) LVDT. (b) Output signals. (c) Angular LVDT.

increase and V_2 decrease, giving an output voltage which is dependent on the displacement, and in phase with the primary voltage. If the core moves in a negative direction, voltage V_1 will decrease and V_2 increase, giving an antiphase output voltage dependent on the displacement. The output signal thus has an

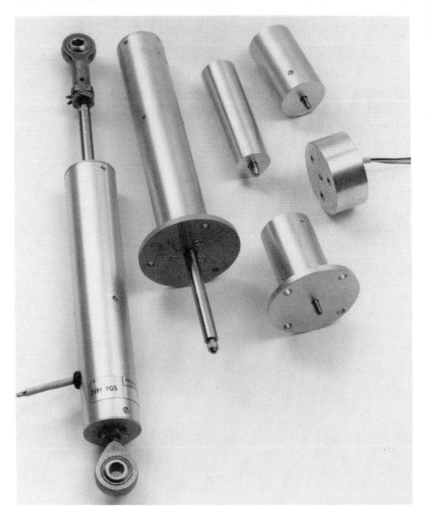

Fig. 4.33 A selection of typical LVDTs. (Photo courtesy of Penny & Giles.)

amplitude indicating the magnitude of the displacement, and phase indicating the direction, as in fig. 4.32b. This AC signal can be converted to a DC signal by means of a phase-sensitive rectifier similar to those described earlier in section 4.3.4 (fig. 4.16).

A related device for small angular measurements is the E transformer of fig. 4.32c. This uses rotation of the pole piece to alter coupling to the two secondaries, giving an output identical to fig. 4.32b.

A commercial device, such as fig. 4.33, is a simpler device to use than fig. 4.32 might imply. It is powered by a DC supply and incorporates its own high-frequency oscillator (typically 10 kHz) and phase-sensitive rectifier. Giving a DC output, and needing a DC supply, it behaves like a potentiometer.

There are many advantages to LVDTs. They have total electrical isolation and are contactless devices, so friction effects and wear are minimal. Unlike potentiometers they have a virtually infinite life and do not go intermittent with age. Less obviously, the coils can be sealed against liquid ingress with the movable core left open. This allows LVDTs to be used in corrosive atmospheres without the need for sealing collars around the spindle. Submersible transducers can operate at depths of over 150 m.

LVDTs are available for a wide range of displacements from 0.1 mm to about 500 mm. They are very stable and have virtually infinite resolution (although the display/control devices connected to them will define some level of resolution themselves). They are, however, more expensive than simple potentiometers.

4.5.3. Variable inductance transducers

The inductance of a coil can be varied either by a movable core, as in fig. 4.34a, or by a movable pole piece on an E core, as in fig. 4.34b. The inductance in both cases is related to the input displacement, and can be measured by an AC bridge or by the current produced by the application of a known AC voltage.

The variable inductance transducer (also known as the variable reluctance transducer) has a unidirectional output and is far less common than the LVDT. Figure 4.34b is, however, the basis of many proximity detectors (see section 4.6).

4.5.4. Variable capacitance transducers

The capacitance of a parallel plate capacitor such as fig. 4.35a is given by:

$$C = \frac{\varepsilon A}{d} \text{ farads} \tag{4.10}$$

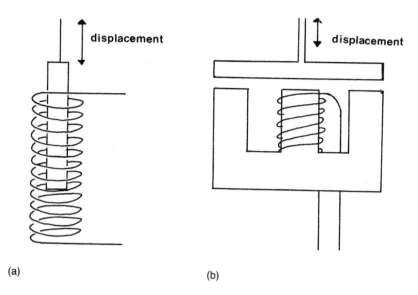

(a)   (b)

Fig. 4.34   Variable inductance displacement transducers. (a) Movable core. (b) E core.

where ε is the permittivity of the material between the plates (8.854 pF m^{-1} for free space), A is the cross-sectional area, and d is the separation.

Each of the terms in equation 4.10 can be varied to provide a displacement transducer. In fig. 4.35b the separation is varied, and in fig. 4.35c the area. An angular version of fig. 4.35c is used as a tuning control on many radio receivers, varying the C of an LC tuned circuit. In fig. 4.35d a dielectric material is moved

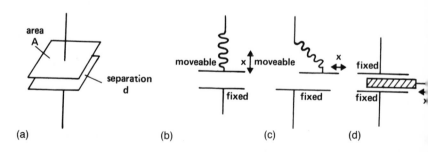

(a)   (b)   (c)   (d)

Fig. 4.35   Fundamentals of variable capacitance displacement transducers. (a) Capacitor basics. (b) Variable separation. (c) Variable area. (d) Variable dielectric.

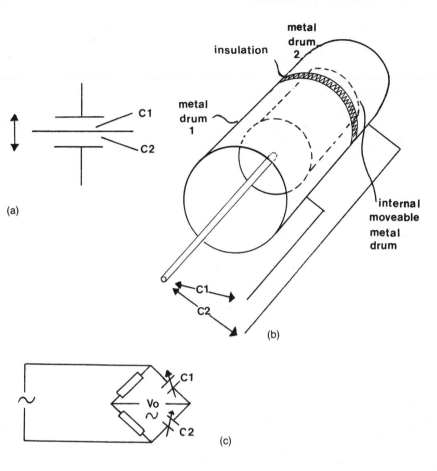

Fig. 4.36 Linearising the response from capacitive displacement transducers. (a) Two-capacitor sensor. (b) Practical transducer. (c) Measuring circuit.

between the plates altering ε. Examination of equation 4.10 shows that figs. 4.35c and d give a linear relationship between capacitance and displacement.

Unfortunately, for practical devices the capacitance and the change given by a displacement is small. A 30 mm square capacitor with a plate separation of 0.5 mm has a capacitance of 16 pF, and this changes by 4 pF for a 0.1 mm displacement (*d* decreasing; the change is 2.7 pF for *d* increasing because of the non-linearity of equation 4.10).

Such changes are difficult to measure directly. Possible methods are via the usual voltage/current relationship for a capacitor, or varying C in an LC tuned oscillator to give a variable frequency. More commonly, however, a differential circuit is used. In fig. 4.36a, the movable centre plate is displaced, causing the capacitance of one side to increase and the other to decrease. Similarly the variable area tubular capacitor of fig. 4.36b causes opposite changes in the two fixed plates. These devices can be connected directly into an AC bridge, as in fig. 4.36c. One particular advantage of this arrangement is that a linear relationship between V_o and displacement is obtained even for fig. 4.36c, as shown below:

$$C_1 = C_0 \frac{d}{d + x} \quad , \qquad C_2 = C_0 \frac{d}{d - x} \qquad (4.11)$$

where C_0 is the zero-position capacitance, x is the displacement, and d is the zero-position separation.

$$V_2 = \frac{(d + x)\, V_i}{(d - x) + (d + x)} = \frac{(d + x)\, V_i}{2d} \qquad (4.12)$$

The voltage at the junction of the resistors is $V_i/2$, so:

$$V_o = V_i \left(\frac{d + x}{2d} - \frac{1}{2} \right) \qquad (4.13)$$

$$= \frac{V_i\, x}{2d} \qquad (4.14)$$

The variable area arrangement gives a similar result.

There are few pure displacement transducers based on the principles above, but many transducers for other process variables use a small displacement as an intermediate stage. Pressure transducers, for example, are often based on fig. 4.36a with the pressure transducer's diaphragm being used as the movable plate.

4.6. Proximity detectors

Sequencing applications often require indication of the state of items such as valves, flow dampers, etc. In many cases, all that is needed is an on/off signal at the limits of travel (e.g. valve fully open/fully shut). Traditionally these applications use mechanical limit switches. These devices, however, are bulky and have a limited life (particularly where the environment is hostile because

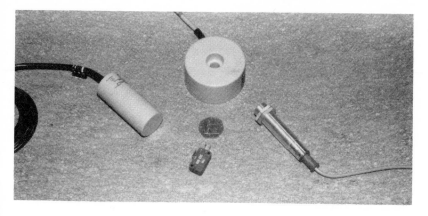

Fig. 4.37   A selection of proximity detectors.

of dust, moisture, temperature or vibration).

A proximity detector can be considered as a solid-state limit switch, and is usually based on the inductive transducer of fig. 4.34b. The transducer contains a coil whose inductance changes as it comes close to a metal surface. This inductance change is detected by a circuit within the transducer, causing the output to switch to some preset level. In a 'limit switch' application, all that is required by way of a 'striker' is a small metal plate that moves in front of the detector at the correct position.

AC-powered devices use two wires and behave just like a switch (apart from a small leakage current in the off state, typically 1 mA). DC-powered devices use three wires: two for the supply and an output which switches between the supplies. DC-powered circuits use a high-frequency internal oscillator which gives a faster response. Detection at 2 kHz is possible with devices such as fig. 4.37.

Sensing distances up to 20 mm are feasible, but 5–10 mm is more common. Maximum sensitivity is obtained with steel or iron, the sensitivity being reduced by over 50% for materials such as brass, copper or aluminium.

Inductive detection can obviously only be used with metallic targets. Capacitive proximity detectors work on the change of capacitance caused by the target, and accordingly work with paper, glass, wood and other materials. Their one major disadvantage is that they need adjustable sensitivity to cope with different applications (unlike inductive versions which are fit-and-forget). Sensing distances up to 50 mm are feasible.

Retro reflective photocells (see section 8.7) are another alternative to inductive detection. Devices superficially similar to fig. 4.37 contain an infrared transmitter and photocell in the head, and detect the target by reflected light. Sensitivity obviously depends on the target surface, but is typically 100 mm for non-rusty mild steel.

4.7. Integration of velocity and acceleration

Velocity is the time integral of acceleration, and displacement the time integral of velocity. Given a transducer for velocity or acceleration, displacement can be obtained by one, or two, integrations. Care must be taken to avoid integration of offset signals. The technique is not widely used in industry, but is the basis of maritime and astronomical inertial navigation systems.

Chapter 5
Flow transducers

5.1. Introduction

The measurement of flow is an essential part of almost every industrial process, and many techniques have evolved. The word 'flow' is a general, and not very precise, term that is used to describe distinctly different quantities.

Volumetric flow is the commonest, and is used to measure the volume of fluid past a given point per unit time (e.g. $m^3 \, s^{-1}$). It may be indicated at the temperature and pressure of the fluid, or *normalised* to some standard temperature and pressure by the relationship:

$$V_n = \frac{P_m V_m T_n}{P_n T_m} \tag{5.1}$$

where V_n is the normalised volumetric flow at pressure P_n and absolute temperature T_n, and V_m is the measured flow at pressure P_m and absolute temperature T_m.

Mass flow is the mass of fluid past a given point per unit time (e.g. $kg \, s^{-1}$). In the case of gases, there is an obvious relationship to normalised volumetric flow as density is dependent on temperature and pressure:

$$M = \varrho_n V_n \tag{5.2}$$

where M is mass flow, V_n is normalised volumetric flow and ϱ_n is the density at the normalised conditions.

Velocity of flow is the velocity with which the fluid is moving past a given point. Care is needed in the measurement because the velocity may not be equal across the pipe or duct. Point velocity of flow measurement is used, for example, in checking flows over car bodies in wind tunnels.

Flow measurement is often as a measurement of some other property. Fiscal measurement for domestic gas, for example, is concerned with charging for the supply of energy. This is achieved by measuring normalised volumetric flow or mass flow and applying an 'energy content' conversion factor measured separately by the gas-supplying authority.

There are several classes into which flow measurement can be divided. An important one is liquids, gases or slurries. Flow measurement in liquids is the simplest because in most cases the liquid can be considered incompressible (thereby simplifying the analysis). With gas flow measurements it is nearly always necessary to make correction for temperature and pressure and to make allowance for the compressibility of the gas. Slurries are liquids with suspended solids and can vary from mud-like substances to relatively clear liquids carrying large pieces of solid matter. The indeterminate nature of the fluid causes measuring difficulties. Another division is closed pipes or open ducts. Most industrial measurement is done in closed pipes, but water and sewage authorities perform measurements in open ducts or conduits. This requires different techniques.

5.2. Differential pressure flowmeters

5.2.1. Basic theory

If a constriction is placed in a pipe through which a fluid is flowing as in fig. 5.1, a differential pressure will be developed across the constriction which is dependent on the volumetric flow. This

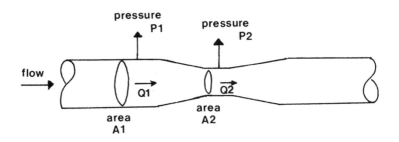

Fig. 5.1 The basis of differential pressure flowmeters; because $Q_1 = Q_2$, the flow velocity must increase in the region of A2 and cause a pressure difference between P1 and P2.

simple principle is the basis of the commonest flow measuring devices, namely orifice plates and venturi tubes. To be used as a flow transducer, however, it is necessary to be able to calculate the volumetric flow from the differential pressure. This is done by considering the energy in the fluid.

The energy in a unit mass of fluid has three components:

(a) Kinetic energy due to its motion, given by $v^2/2$ where v is the velocity.

(b) Potential energy gh where g is the acceleration due to gravity and h is the height above some datum.

(c) Energy due to the pressure of the fluid (often called, rather confusingly, flow energy). This is similar to potential energy, and is given by P/ϱ where P is the pressure and ϱ the density at the fluid's temperature and pressure.

The energy of a unit mass of liquid at any point is the sum of these three components, and because there is no net gain or loss of energy across the constriction we can say:

$$\frac{v_1^2}{2} + gh_1 + \frac{P_1}{\varrho_1} = \frac{v_2^2}{\varrho_2} + gh_2 + \frac{P_2}{\varrho_2} \tag{5.3}$$

This is the fundamental equation for differential pressure flowmeters.

5.2.2. Turbulent flow and Reynolds number

If equation 5.3 is to be used as the basis for flow measurement, there is an underlying assumption that the velocity of the fluid is the same at all points across the pipe. At low flows this may not be so because of frictional effects at the pipe wall. This leads to a velocity profile similar to fig. 5.2a which is called streamline, or laminar, flow. If the flow velocity increases, turbulence starts and at a sufficiently high flow the velocity profile becomes equal across the pipe, as shown in fig. 5.2b. In general, differential pressure flowmeters can only be used where the flow is turbulent.

For a given fluid, the flow condition is indicated by its Reynolds number, Re, given by:

$$Re = \frac{vD\varrho}{\eta} \tag{5.4}$$

where v is the fluid velocity, D is the pipe diameter, ϱ is the fluid density, and η is the fluid viscosity. (The kinematic viscosity, η/ϱ, is sometimes given in the literature rather than individual values of ϱ and η.) The Reynolds number is a ratio, and hence dimensionless. The larger the value of Re, the more turbulent the flow. In general, if $Re<2000$ the flow is laminar. If $Re>10^5$ the flow is fully turbulent.

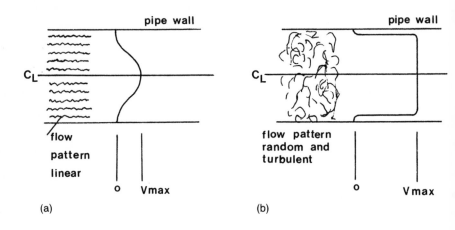

Fig. 5.2 The effect of flow velocity on flow pattern and velocity profile. (a) Streamline (laminar) flow. Velocity profile is not constant if flow stagnates at walls. (b) Turbulent flow. Velocity profile is uniform across pipe.

5.2.3. Incompressible fluids

If we consider an incompressible fluid in fig. 5.1 and equation 5.3, $\varrho_1 = \varrho_2 = \varrho$. For simplification, we will also assume that the pipe is horizontal so $gh_1 = gh_2$. Equation 5.3 simplifies to:

$$\frac{v_2^2 - v_1^2}{2} = \frac{P_1 - P_2}{\varrho} \tag{5.5}$$

The ingoing flow, Q_1, is A_1v_1, and the flow at the constriction Q_2 is A_2v_2. Because flow is conserved, $Q_1 = Q_2 = Q$ and $v_1 = Q/A_1$, $v_2 = Q/A_2$. Substituting into equation 5.5 gives:

$$Q = \frac{A_2}{\sqrt{1 - (A_2/A_1)^2}} \sqrt{\frac{2(P_1 - P_2)}{\varrho}} \tag{5.6}$$

In practice, this equation needs modification because A_1, A_2 do not correspond exactly to the pipe area and the area of the constriction. In the case of an orifice plate, fig. 5.3, the minimum area (called the plane of vena contracta) occurs downstream of the constriction.

A more practical equation is:

$$Q = C_D \frac{A_2}{\sqrt{1 - (A_2/A_1)^2}} \sqrt{\frac{2(P_1 - P_2)}{\varrho}} \tag{5.7}$$

where C_D is called the discharge coefficient, and is effectively a 'frig factor'. Typical values are 0.97 for a venturi tube and 0.6 for orifice plates.

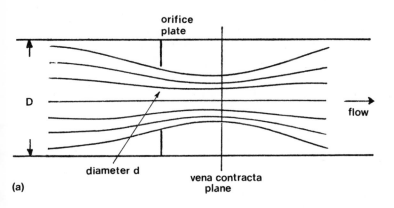

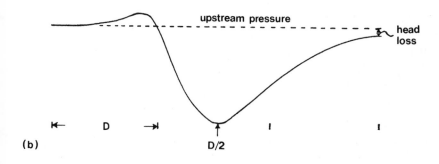

Fig. 5.3 Pressure curve for an orifice plate. (a) Orifice plate. (b) Pressure curve.

In practical calculations it is easier to work with diameters of the pipe (D) and the restriction (d). If we define m as the area ratio A_2/A_1 ($m = d^2/D^2$) and E as the velocity of approach factor $1/\sqrt{(1 - m^2)}$, equation 5.7 can be written in a more convenient form.

$$Q = C_D E \frac{\pi d^2}{4} \sqrt{\frac{2(P_1 - P_2)}{\varrho}} \qquad (5.8)$$

This equation is used for the design of orifice plates. The user can simplify it further to:

$$Q = K\sqrt{\Delta p} \qquad (5.9)$$

where Δp is the observed differential pressure and K is a constant for a specific application. Note the non-linear square-root relationship.

The difficulty in using equation 5.8 is establishing a value for C_D. This can be established by experiment with independent flow measurement. Alternatively British Standards Institute BS 1042, Part 1: 1964, Methods for the Measurement of Fluid Flow in Pipes, Orifice Plates, Nozzles and Venturi Tubes, provides graphs from which C_D can be calculated. A mathematical approach (the Stolz equation) is also given in ISO 5167:1980.

5.2.4. Compressible fluids

With gases, the fluid density does not remain constant through the constriction, so $\varrho_1 \neq \varrho_2$ in equation 5.3. It is also necessary to consider mass equality rather than volumetric flow equality through the constriction. These changes complicate the calculations for gas flowmeters.

The equation 5.6 is modified to:

$$Q_m = C_D \varepsilon \frac{A_2}{\sqrt{1 - (A_2/A_1)^2}} \sqrt{2\varrho_1(P_1 - P_2)} \qquad (5.10)$$

where ϱ_1 is the density at pressure P_1 (upstream) and ε is defined as the expansion ratio or expansibility factor. C_D is the discharge coefficient factor defined earlier. Note that Q_m is *mass* flow rate.

Calculation of ε is complex as it depends on γ (the ratio of specific heats C_p/C_v), P_2, P_1, A_2, A_1. These all interact so it is not simply a matter of plugging values into an equation. BS 1042 uses a nomograph to calculate ε (unfortunately in imperial units). ISO

5167 gives an equation which is used repetitively for successive approximations of ε until a value is obtained to the required accuracy. This approach, called regression, is well suited to computer analysis but rather laborious for hand calculation.

A practical version of equation 5.10 (BS 1042) is:

$$Q_m = \varepsilon E C_D A_2 \sqrt{2\varrho_1(P_1 - P_2)} \tag{5.11}$$

where E is the velocity of approach factor defined earlier.

Note that the upstream pressure is effectively included in equations 5.10 and 5.11 via ϱ_1. A gas differential pressure flowmeter is calibrated for one particular upstream pressure.

5.2.5. Orifice plates

An orifice plate is used to make an abrupt change in the pipe area, and simply consists of a circular plate usually inserted between pipe flanges as shown in fig. 5.4. This produces a pressure differential which is usually measured at D upstream and $D/2$ downstream where D is the pipe diameter.

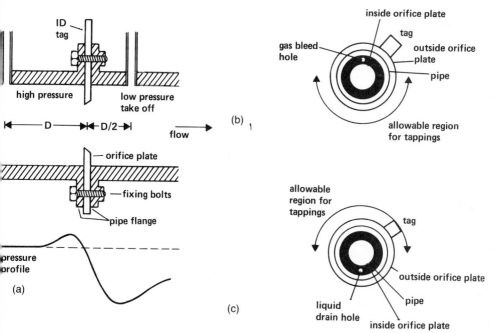

Fig. 5.4 Flow measurement with orifice plates. (a) Orifice plate arrangement. (b) Liquids. (c) Gases.

Figures 5.3 and 5.4 also show that the final downstream pressure is lower than the upstream pressure; the orifice plate has caused a permanent loss of pressure called the head loss. This can be as high as 50% of the upstream pressure. In applications where this cannot be tolerated, a venturi tube (described in section 5.2.6) is used.

There is more to manufacturing an orifice plate than drilling a hole in a circular plate. There will be a considerable force on the plate, which must be sufficiently rigid to resist distortion. BS 1042 recommends a *maximum* thickness of $0.1D$. The upstream edge of the plate must have a sharp edge as shown in fig. 5.4 (which implies that an orifice plate has an upstream and downstream face which must be identified). The edge will suffer abrasive effects from the fluid flow, and should be constructed of materials such as stainless steel to avoid excessive wear.

A small hole must be drilled in the plate as shown in figs. 5.4b and c. For liquids, the hole should be vertically above the opening to allow passage of trapped air or gas. For gases or vapours the hole should be below the opening, flush with the pipe wall, to allow condensate to pass. To avoid errors, the pressure tappings must be displaced by at least 90° from the hole as shown. The identification tag is essential for future users of the system. This should show its identification in the system (e.g. FE207) and the internal diameter.

Care must be taken when installing orifice plates in pipe runs. Close proximity of bends and control valves can cause local pressure variations, so clear pipe runs of at least $10D$ are required both upstream and downstream of the orifice plate. This can be relaxed if straightening vanes similar to fig. 5.5 are used.

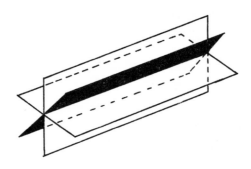

Fig. 5.5 Flow-straightening vanes.

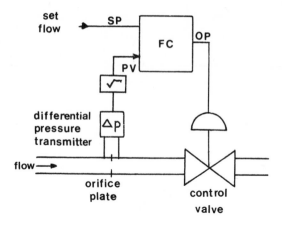

Fig. 5.6 The use of an orifice plate in a flow control loop.

When a gas flow measuring orifice plate is used in conjunction with a flow control valve, as in fig. 5.6, the orifice plate must be situated upstream of the valve. Downstream of the valve the fluid pressure will vary considerably according to the setting of the control valve, causing significant errors via changes of the density, ϱ, in equation 5.11.

There are various arrangements of pressure tappings used with orifice plates. The commonest of these are shown on fig. 5.7. The $D - D/2$ arrangement (sometimes called radius or throat taps) approximate to the theoretical conditions of fig. 5.4. Plate taps and carrier rings are complete assemblies and are used where it is not feasible to drill the pipe or the flange assembly. The nozzle arrangement of fig. 5.7f has a more predictable discharge coefficient and lower head loss but is obviously more expensive to manufacture.

Orifice plates are used in many applications and are probably the commonest flow measuring device. Typical installations are shown in fig. 5.8.

5.2.6. Dall tubes and venturi tubes

The orifice plate produces a large head loss. If this is unacceptable, a smoother obstruction must be used. The two commonest devices are the venturi tube (fig. 5.9a), which is a manufactured assembly, and the Dall tube (fig. 5.9b), which is effectively an insert in a pipe section.

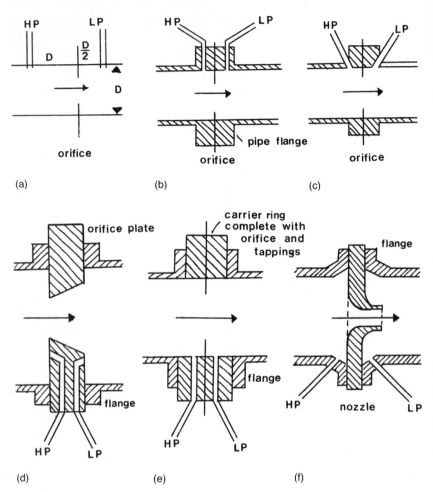

Fig. 5.7 Various installation arrangements for orifice plates. (a) D–D/2 tapping (very common). (b) Flange taps, normally used on large pipes with substantial flanges. Tappings typically ± 25 mm from orifice. (c) Corner taps, drilled obliquely through flange. (d) Plate taps (horizontal scale exaggerated). (e) Orifice carrier (can be factory made and needs no site drilling). (f) Nozzle (gives lower head loss).

The smoother transition lowers the head loss to between 5 and 10% (compared with around 50% for an orifice plate) but the differential pressure is reduced. The output from a venturi tube is about one-third that of an orifice plate operating under similar conditions. The output from a Dall tube is midway between a venturi and an orifice plate.

The venturi and Dall tubes are superior to orifice plates from a

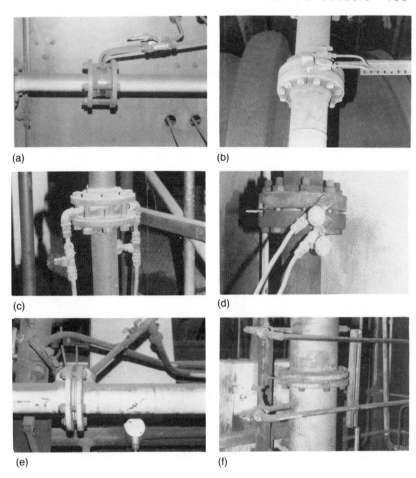

Fig. 5.8 A selection of orifice plates; note tappings & details such as Tag ID. (a) Carrier Assembly. (b) Flange Tapping. (c) Carrier Assembly. (d) Flange Tapping. (e) Corner Tapping. (f) D—D/2.

theoretical instrumentation viewpoint, but have practical drawbacks. They are obviously expensive to manufacture, and difficult to fit. A venturi for a 25 cm pipe with 15 cm throat would be over 2 m in length, and would need at least 1 m free of obstruction or bends up- and downstream of the device.

The flow equations derived for orifice plates apply equally to venturi and Dall tubes. Typical values for C_D are in the range 0.94 to 0.97.

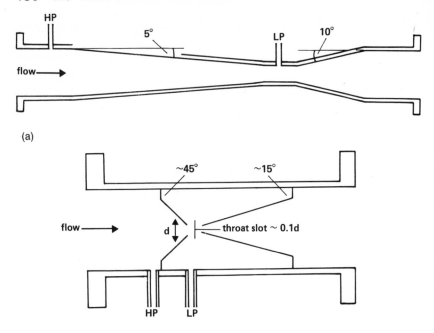

(a)

(b)

(c)

Fig. 5.9 Low head loss primary sensors. (a) Venturi tube (note the length!). (b) Dall tube. (c) Photograph of Dall tube (Photo courtesy of Kent Industrial Measurements.)

5.2.7.   The Pitot tube

Previous differential pressure devices have measured volumetric or mass flow through a pipe. The Pitot tube, shown in its simplest form in fig. 5.10, measures flow velocity at one point. It is used for velocity measurements in, say, car and aircraft body testing in wind tunnels, and, slightly modified, as an air speed indicator.

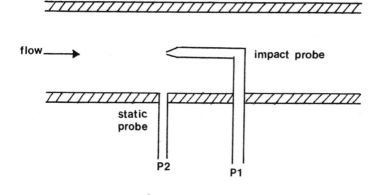

Fig. 5.10   Pitot tube

At the tip of the impact probe, the fluid is brought to rest. This leads to a rise in pressure at P_1 with respect to P_2, which can be evaluated by considering the flow and kinetic energy at the impact and static probe.

Assuming an incompressible fluid, and no difference in height between the static and impact probe, the energy per unit mass is calculated:

(a) at the impact probe P_1/ϱ since there is no kinetic energy;
(b) at the static probe $P_2/\varrho + V^2/2$.

These will be equal so:

$$\frac{P_1}{\varrho} = \frac{P_2}{\varrho} + \frac{V^2}{2}$$

or

$$V = \sqrt{\frac{2(P_1 - P_2)}{\varrho}} \tag{5.12}$$

The relationship for a compressible fluid is, naturally, more complex, but rather surprisingly the effect is small. Equation 5.12 can be used for gases at velocities up to 150 m s^{-1} with an error of about 2%.

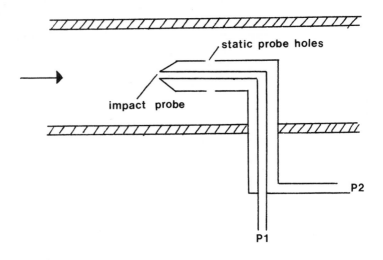

Fig. 5.11 Practical insertion Pitot tube.

A practical Pitot tube, shown in fig. 5.11, consists of two concentric tubes. The inner, open at one end, acts as the impact probe and the outer as the static probe via holes around its circumference. Because of the small radial separation of the probes, the measurement is unaffected by the type of flow (turbulent or laminar) in the pipe. Alignment is important; a 20° offset gives about 2% error.

Pitot tubes can be used to infer volumetric flow if the flow is turbulent, or an averaging tube can be employed (such as the Annubar of fig. 5.12). (The name Annubar is a registered trade mark of the Dietrich Standard Corporation.) Figure 5.13 shows a typical Annubar installation. An alternative approach uses a single Pitot tube which is traversed across the pipe at regular intervals by an electric motor to enable an average flow to be computed.

The lack of complications such as discharge coefficients and expansibility factors and a negligible head loss makes the Pitot tube an attractive sensor. Its main disadvantage is a low differential pressure which is often at the lower limit of

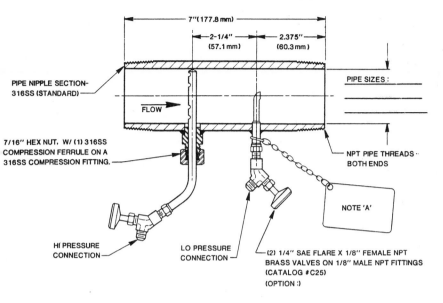

Fig. 5.12 Annubar pipe insert. (Annubar is a trade name of the Dietrich Corporation; illustration courtesy of their UK representatives, Auxitrol Ltd.)

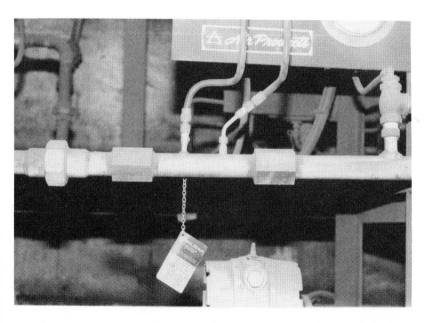

Fig. 5.13 An Annubar installation (oxygen flow) by Auxitrol Ltd & Air Products.

availability of industrial pressure transducers. The small diameter of the impact tube also makes it vulnerable to blockage from fluid-borne particles.

The design of Pitot tubes is covered in the BSI publication BS 1042, Part 2: 1943.

5.2.8. Measurement of differential pressure

Conversion of the diffential pressure to an electrical signal requires a differential pressure (Δp) transducer. The construction of these devices is described in chapter 3; this section describes practical details. Orifice plate connections are shown; piping for venturi or Pitot tubes is similar.

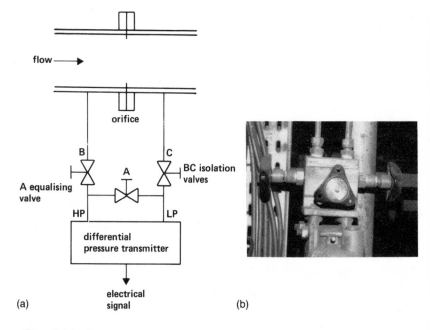

Fig. 5.14 Connection of differential pressure flow sensor to Delta P transmitter. (a) Valve arrangement. (b) Manifold for differential pressure transmitter.

The Δp transmitter is connected via a manifold block as shown in fig. 5.14. This allows the transmitter to be isolated for removal or maintenance by means of the valves B and C. Valve A is an equalising valve and is used during zero adjustments when B and

C are closed and A is open. During normal operation A is closed. When these valves are being operated, care must be taken to ensure that full pipe pressure does not get 'locked in' one leg (which could damage the diaphragm in the transmitter). The order should always be: open A, close B, C; or open B, C, close A.

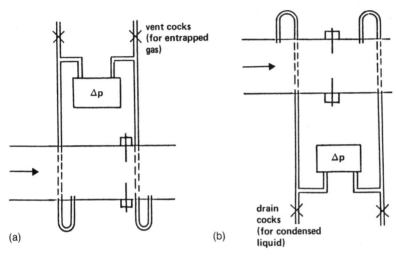

Fig. 5.15 Preventing build up of liquid sumps and gas pockets if straight pipe connections are not feasible. (a) Liquid flow. (b) Gas flow.

Care must be taken with the piping installation to avoid problems from condensed liquids (with gas measurement) or air pockets (with liquid measurement). Ideally a gas measurement system should have the Δp transmitter above the orifice plate with straight pipe drops (so the pipes drain). Similarly liquid measurement systems should have the Δp transmitter below the orifice plate with straight pipe rises (so air bubbles rise back into the pipe).

If these ideal arrangements are not possible, the piping should be arranged as in fig. 5.15 with vent/drain cocks. In the liquid arrangement of fig. 5.15a the line pressure must be such that the Δp pipes fill with the vent cocks open. If both Δp pipes are not full, there will be errors due to the different liquid heads in the two pipes.

The square-law relationship of differential pressure flowmeters necessitates a square-root extraction somewhere in the installation. Usually this is performed electronically on the Δp

transmitter output; either via a multibreak point Op Amp circuit (see chapter 1 in Volume 2) or digitally via look-up tables.

The square-root extraction determines the lowest measurable flow, as a small zero offset makes for a large error at low flow. The effect of the square root can be seen in the accompanying table which shows (as a percentage of full scale) true flow, true Δp, Δp with a +0.5% zero error and indicated flow.

Flow	Δp	Δp + offset	Indicated flow
100	100	100.50	100.25
75	56.25	56.75	75.33
50	25.0	25.5	50.5
25	6.25	6.75	25.98
10	1.0	1.5	12.25
5	0.25	0.75	8.66
2	0.04	0.54	7.35
1	0.01	0.51	7.14

With a −0.5% zero error, the indicated flow will be zero at a true flow of about 7%. On a 4–20 mA signal, 0.5% corresponds to a zero error of less than 0.1 mA. Zero offset is not just a setting-up problem; zero drift also occurs as a result of hysteresis, temperature effects and ageing.

This low flow error limits differential pressure flow measurement to a turndown ratio (high flow/low flow) of at most 6 : 1. If a greater range is required, two or more Δp transmitters are used with range switching between them.

5.3. Turbine flowmeters

As its name suggests, a turbine flowmeter consists of a small turbine (usually four-bladed) placed in the flow as shown in fig. 5.16. Within a specified flow range (usually about 10 : 1 turndown) the speed of rotation is directly proportional to flow velocity.

The turbine blades are constructed of ferromagnetic material and pass beneath a magnetic detector operating as a variable reluctance transducer, which produces an output voltage approximating to a sine wave of the form:

$$E = A\omega \sin N\omega t \qquad (5.13)$$

where A is a constant, ω is the angular velocity (proportional to flow velocity) and N is the number of blades. Both the output amplitude and frequency are proportional to flow velocity, although frequency-dependent circuits are normally employed to give a flow-dependent current or voltage output.

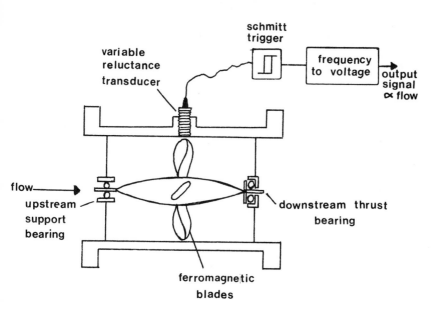

Fig. 5.16   Turbine flowmeter.

The lower flow limit is determined by frictional effects on the rotor or an unacceptably low amplitude sine wave from the magnetic detector. Other non-linearities occur from magnetic drag from the pick-up and viscous drag from the fluid itself. A typical accuracy is about ±0.5% with a 10 : 1 turndown.

The flowmeter can be affected by swirling of the fluid itself. This can be overcome by the use of straightening vanes (fig. 5.5) in the pipe upstream of the flowmeter.

Turbine flowmeters are relatively expensive, and less robust than differential pressure devices. Having mechanical rotating parts, they are prone to damage from suspended solids. Their main advantages are a linear output, superior turndown and a pulse signal that can be used directly for flow totalisation. Figure 5.17 shows a typical turbine flowmeter.

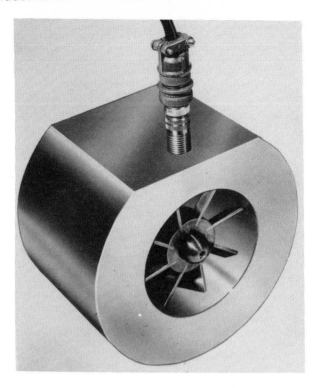

Fig. 5.17 Turbine flowmeter. (Photo courtesy of Kent Industrial Measurements.)

5.4. Variable area flowmeter

A variable area flowmeter consists of a tapered float in a vertical tapered glass tube as shown in fig. 5.18. Fluid flows vertically past the float, which rises in the tube to a position which is dependent on the volumetric flow. The float forms an obstruction in the pipe, with cross-section as in fig. 5.18a. There is therefore a pressure drop across the float Δp which will produce an upward force $\Delta p.A_1$. There will also be a downward force due to the displacement of the float (Archimedes' principle). If the float is in equilibrium, these forces will balance, i.e.

$$\Delta p.A_1 + V\varrho_2 g = V\varrho_1 g \qquad (5.14)$$

where V is the volume of the float, ϱ_2 is the density of the fluid, and ϱ_1 is the density of the float. This can be rewritten

$$\Delta p = h(\varrho_1 - \varrho_2)g \qquad (5.15)$$

where h is the float thickness.

From equation 5.10 we have:

$$Q = KA_2\sqrt{\Delta p} \tag{5.16}$$

where A_2 is the area of the annulus and K is a constant.

From equation 5.15, the float will move to maintain a constant differential pressure Δp. From equation 5.16, with constant Δp, the area of the annulus, A_2, is proportional to the volumetric flow.

The measured variable is the float position, x. This is related to A_2 by fig. 5.18c, where θ is the angle of the tube. θ is very small, typically a few degrees, so we can write:

$$A_2 = 2\pi dx \tan\theta \tag{5.17}$$

Substituting back into 5.16 gives:

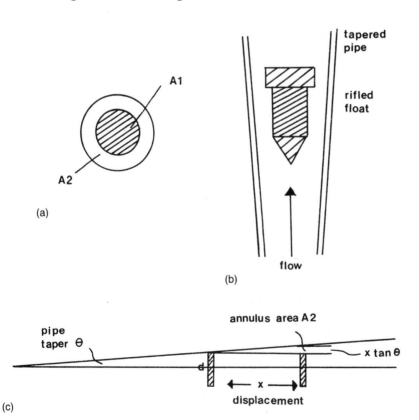

Fig. 5.18 Variable area flowmeter. (a) Cross-section of float and pipe. (b) Side view. (c) Trigonometrical relationships.

$$Q = 2\pi K dx \tan \theta \sqrt{\Delta p} \tag{5.18}$$

or

$$Q = Bx \tag{5.19}$$

where B is a constant. The displacement, x, is directly proportional to flow.

Fig. 5.19   Variable area float flowmeter.

A practical device is shown in fig. 5.19. The float has a slight rifle around its perimeter to induce a stabilising spin. The variable area flowmeter is simple, linear and cheap, and has a good turndown ratio of about 10 : 1. It cannot be easily adapted for remote indication (although devices using the float as the core of an LVDT are available). Obviously, the fluid must be a clean and clear gas or liquid. If the float is made of a ferromagnetic material, a magnetic detector placed alongside the tube can be used to give a low (or high) flow alarm.

5.5.   Vortex shedding flowmeter

If a bluff (non-streamlined) body is placed in a pipe as in fig. 5.20, the flow cannot follow the surface of the object, and vortices

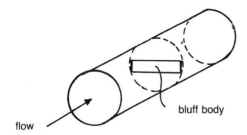

(a)

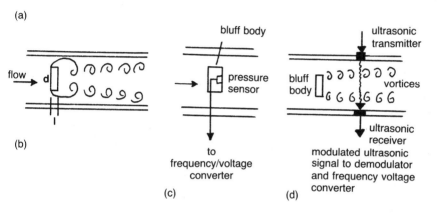

(b)

(c)

(d)

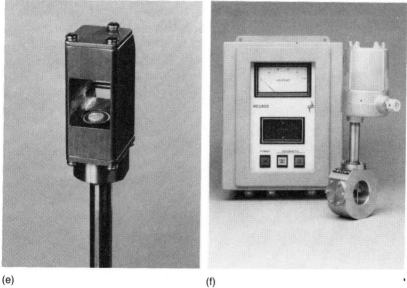

(e)   (f)

Fig. 5.20 Vortex shedding flowmeter. (a) Construction. (b) Vortex formation. (c) Pressure sensor on downstream face of bluff body. (d) Ultrasonic method. (e) Vortex shedding head. (f) Complete unit. (Photos courtesy of Scheme Engineering.)

detach themselves from the downstream side. The frequency of this vortex shedding is directly proportional to the volumetric flow rate, i.e.:

$$Q = Kf \qquad (5.20)$$

where K is a constant, Q is the volumetric flow and f is the observed frequency. K is effectively constant for flows with Reynolds numbers in excess of 10^3. Surprisingly, K is largely independent of fluid density and viscosity, and is determined by the pipe diameter D, the obstruction diameter d, and the obstruction shape (e.g. rectangular, circular). A rectangular cross-section with $d/D = 0.25$ and $d/l = 1.5$ is usually used. The vortex shedding frequency is typically a few hundred hertz.

The vortex shedding flowmeter is an attractive device. It has no moving parts, low head loss (compared with an orifice plate), an ability to work to relatively low Reynolds numbers, a linear response, and a turndown ratio of about 15 : 1. Its one drawback is that secondary sensors are needed to detect the vortices.

The vortices manifest themselves as sinusoidal pressure variations. These can be detected by a sensitive diaphragm on the trailing edge of the bluff body. The movement of the diaphragm can be converted to a sinusoidal electrical signal by capacitance variations or via strain gauges. Alternatively, an ultrasonic beam across the pipe downstream of the bluff body can be modulated by the vortices. Whatever the detection method, the resulting sinusoidal output is converted to a constant-amplitude square wave prior to conversion to a linear voltage or 4–20 mA current signal by a frequency-to-voltage converter.

5.6. Electromagnetic flowmeter

In fig. 5.21a, a conductor length l is moving with velocity perpendicular to a magnetic field of flux density B. By Faraday's law of electromagnetic induction, a voltage E is induced where:

$$E = Blv \qquad (5.21)$$

The electromagnetic flowmeter is based on the above equation, with the fluid forming the conductor. In fig. 5.21b, a conducting fluid is passing down a pipe with mean velocity $\bar{v}$ (the use of mean velocity covers both laminar and turbulent flow conditions). An insulated section is inserted in the pipe as shown, and a magnetic field B is applied perpendicular to the flow. Two electrodes are inserted through the insulated section into the fluid, to form, with the fluid, a moving conductor of length D where D is the pipe diameter. A voltage E will occur across the electrodes given by:

$$E = BD\bar{v} \qquad (5.22)$$

The electromagnetic flowmeter therefore measures average flow velocity.

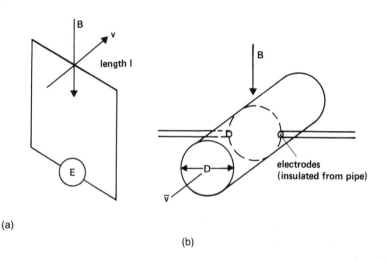

(a)

(b)

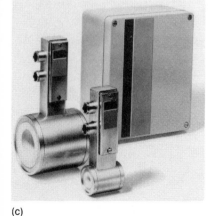

(c)

(d)

Fig. 5.21 Electromagnetic flowmeter. (a) Electromagnetic principles. (b) Schematic. (c) Electromagnetic flowmeters. Note insulated PTFE lining. (Magflo unit, photo courtesy of Danfoss Ltd.). (d) Electrodes being installed after lining of sensor. (Photo courtesy of Danfoss Ltd.)

Figure 5.21 implies a steady field and a DC voltage. In practice, an AC field (and hence an AC-induced voltage) is used to prevent electrolysis effects at the electrodes, and also to minimise errors from DC voltages arising from thermoelectric and electrochemical effects which are of the same order of magnitude as the induced voltage. There is also the possibility that a DC current might actually distort the flow (magnetohydrodynamic effects).

The electromagnetic flowmeter is a flow velocity measuring device, but in general volumetric flow will be proportional to average velocity providing the flow characteristic (laminar/turbulent) does not change.

Although the electromagnetic flowmeter is linear, and has a good turndown ratio of about 15 : 1 and effectively zero head loss, it has several disadvantages. It is bulky and expensive to install (both the instrument itself and the physical installation/cabling) and can only be used on fluids with a conductivity in excess of 1 mS m^{-1}. This excludes all gases and many liquids. The main use of the device is in the flow measurement of 'difficult' liquids such as slurries with a large solids content.

5.7. Ultrasonic flowmeters

5.7.1. Doppler flowmeter

The Doppler effect occurs when there is a relative motion between a sound transmitter and receiver as shown in fig. 5.22a. If the transmitted frequency is f_t Hz, V_s the velocity of sound, and V_r the relative velocity, then the observed frequency f_r is given by:

$$f_r = f_t \frac{(V_r + V_s)}{V_r} \qquad (5.23)$$

i.e. the pitch of the note rises. If the transmitter and receiver are moving apart with relative velocity V_r,

$$f_r = f_t \frac{(V_r - V_s)}{V_r} \qquad (5.24)$$

i.e. the pitch falls. The effect is commonly heard as a change in the pitch of a car engine as it passes. Equations 5.23 and 5.24

apply for both a stationary transmitter/moving receiver and a moving transmitter/stationary receiver.

Figure 5.22a assumes sound waves, although Doppler shift also occurs in electromagnetic radiation. Doppler shift of microwaves is used for speed measurement in radar 'speed traps', and Doppler shift of visible light (red shift) is used in astronomy to measure the velocity of recession of other galaxies.

A Doppler flowmeter injects an ultrasonic (high frequency, typically a few hundred kilohertz) sound wave into a moving fluid as shown in fig. 5.22c. A small part of this sound wave is reflected off solid matter, vapour and air bubbles or eddies/vortices back to a receiver mounted alongside the transmitter. As it passes through the fluid, the frequency is subject to two changes (one travelling upstream against the flow, and one downstream with the flow).

The received frequency is therefore:

$$f_r = f_t \frac{(v_t + V \cos \theta)}{(v_t - V \cos \theta)} \tag{5.25}$$

where v is the average fluid velocity and θ the angle of the ultrasonic beam to the fluid flow. If Δf is the change in frequency $(f_r - f_t)$ and $V \ll v_t$, equation 5.25 can be rewritten

$$\Delta f = \frac{2f_t}{v_t} V \cos \theta \tag{5.26}$$

The change in frequency is proportional to the average fluid velocity.

Figure 5.22b is a block diagram of a complete flowmeter. The transmitted and received frequencies are added together. This produces an amplitude-modulated signal with an envelope frequency of $\Delta f/2$. This is rectified and smoothed to produce a sinusoidal signal of frequency $\Delta f/2$, which is converted to a DC voltage or current proportional to the average flow velocity.

The Doppler flowmeter is linear, and can be installed without the need to break into the pipe (it is the only real 'clip-on' flowmeter). It can be used with all fluids, and is well suited for use with difficult fluids such as corrosive liquids or heavy slurries. The flowmeter works with laminar or turbulent flow (and even with transitional cases). It is rather expensive and over complex for straightforward flow measurement applications, however.

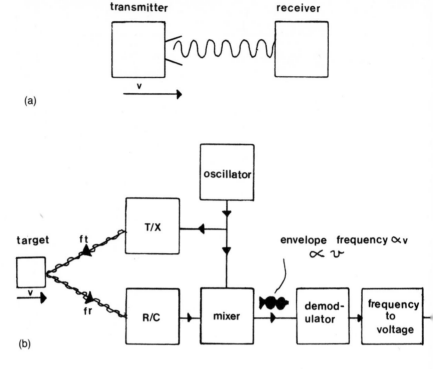

(a)

(b)

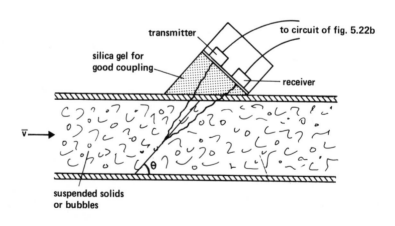

(c)

Fig. 5.22 Ultrasonic flowmeters. (a) Doppler shift principle. (b) Block diagram of Doppler velocity measurement. (c) Practical clip-on ultrasonic flowmeter. (d) Ultrasonic flowmeters. (Photo courtesy of Danfoss Ltd.)

(d)

Fig. 5.22 contd.

5.7.2. Cross-correlation flowmeter

There are many difficulties in measuring the flow of slurries: density changes, abrasion, blockage of sensors, etc. The cross-correlation flowmeter is designed to use the properties of suspended solids to measure the flow.

In fig. 5.23, a slurry is flowing down a pipe from left to right with average velocity V. Two ultrasonic beams, a distance X apart, are passed across the pipe. Random variations in the phase and amplitude of the received signals will occur due to the random distribution of the solids in the fluid. If the distance X is not too large, output 2 will be related to output 1 but delayed by a time T_d which is given by:

$$T_d = X/V \tag{5.27}$$

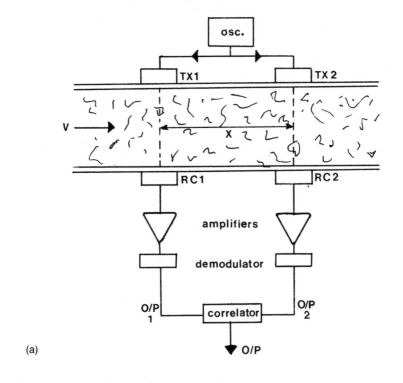

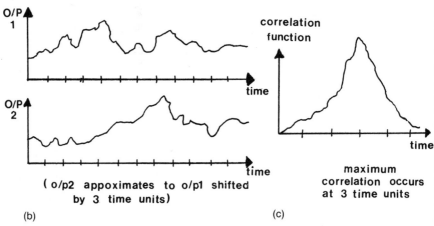

(b)   (c)

Fig. 5.23 Cross-correlation flowmeter. (a) Transducer schematic. (b) Output signals. (c) Correlation function. (d) Cross-correlation flowmeter and electronics rack. (Photo courtesy of Kent Industrial Measurements.)

Obviously the two outputs will not be identical, but with the correct choice of X (usually one pipe diameter) a correlation circuit can determine the value of the delay T_d which gives the

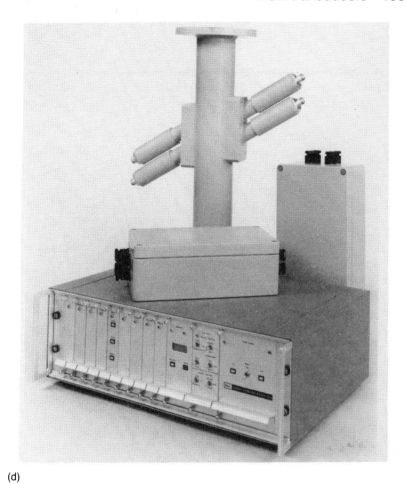

(d)

Fig. 5.23   contd.

maximum on the cross-correlation function as shown on fig.
5.23c. With T_d known, the flow velocity can be deduced from
equation 5.27.

Although ultrasonics are most commonly used with
cross-correlation flowmeters, other techniques are available.
Random variations of conductivity, pressure and even
temperature have all been the basis of successful cross-correlation
flowmeters. Cross-correlation techniques have also been
successfully used for linear measurement of speed with two retro
reflected infrared beams. A typical application is the

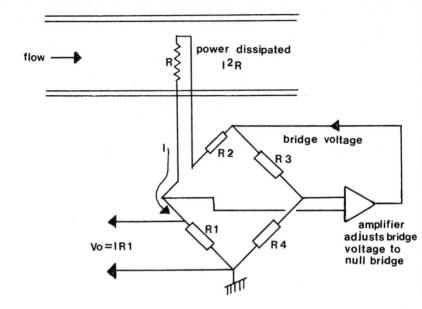

Fig. 5.24   The hot-wire anenometer.

measurement of the velocity of hot steel strip in a high-speed rolling mill.

5.8.   Hot-wire anenometer

If a fluid passes over a hot object, heat is removed. It can be shown that the power loss is given by:

$$P = A + B\sqrt{v}   \qquad (5.28)$$

where v is the flow velocity and A, B are constants. A is related to radiation and B to conduction power loss.

Figure 5.24 shows a flowmeter based on equation 5.28. A hot wire is inserted in the fluid, and maintained at a constant temperature by the self-balancing bridge. If the flow increases, say, more heat is removed, the temperature of the wire falls and its resistance falls. This unbalances the bridge, which is detected and the bridge voltage is increased until the temperature of the wire is restored.

With a constant wire temperature, the heat dissipated by the wire is equal to the power loss, i.e.

$$I^2R = A + B\sqrt{v} \qquad (5.29)$$

or

$$v = K(I^2R - A)^2 \qquad (5.30)$$

where K is a constant.

The current I is measured (or converted to a voltage by the lower resistor in the bridge) from which the flow velocity can be deduced. Obviously the relationship is non-linear, and compensation for changes in fluid temperature is required.

5.9. Injection flow measurement

In some applications, an ad hoc flow measurement is needed where there is no real knowledge of the required range of flow. Figure 5.25 shows a method which can be readily adapted to measure gaseous flow over a wide flow range without a previous detailed study of the system.

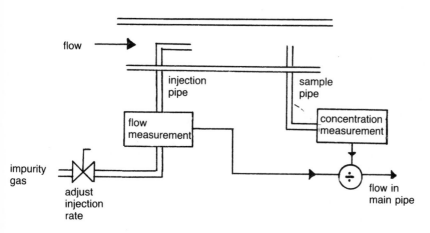

Fig. 5.25 Injection-flow measurement.

An impurity gas is bled into the pipe at a known rate. Typical injection gases are CO_2 for town gas flow measurement, and oxygen for air (the oxygen content of air being fixed, it is easy to detect an increase). A gas analyser is used to sample the percentage of the impurity gas downstream of the injection point (at a point where mixing is complete). As the injection flow rate, and the resulting percentage impurity are known, it is a simple calculation to find the main gas mass flow rate.

5.10   Flow in open channels

Measurement of flow in open channels is required in municipal water and sewage operations. This is most commonly performed by the use of weirs (see fig. 5.26) to produce a fluid head which is dependent on the volumetric flow. The fluid head can be measured by a float system or by measuring the pressure head via a pressure pipe.

The relationships are non-linear. For the rectangular and Cipoletti weir the relationship is:

$$Q = K H^{3/2} \qquad\qquad (5.31)$$

where K is a constant dependent on the weir dimensions. For the V-notch (or Thomson) weir, the relationship is:

$$Q = K H^{5/2} \qquad\qquad (5.32)$$

In all cases, the velocity of the fluid is kept low, or corrections must be made for the kinetic energy of the flow. Design of open-channel flowmeters is covered in BS 3680.

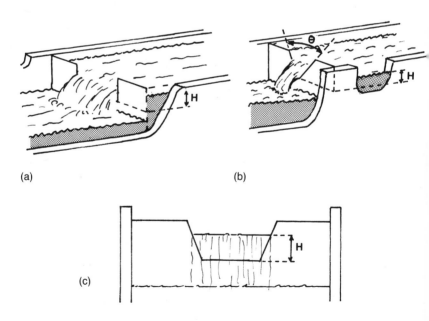

(a)   (b)

(c)

Fig. 5.26 Flow in open channels. (a) Rectangular weir. (b) V notch or Thompson weir. (c) Trapezoid or Cipoletti weir.

Chapter 6
Strain gauges, loadcells and weighing

6.1. First principles

6.1.1. Introduction

Measurement of weight is an essential part of most industrial processes, particularly where batch manufacturing is used. Weight is also the method of determining the value of most goods, so there is often an economic and legal requirement for accurate weight measurement. This chapter is concerned with techniques for measuring weight, and the closely related topic of strain measurement.

There are essentially two techniques of weight measurement,

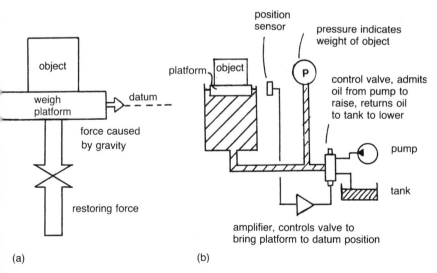

Fig. 6.1 Force balance (null balance) weigher. (a) Principle. (b) Hydraulic implementation.

shown diagrammatically in figs. 6.1 and 6.2. In a null balance weigher the applied weight causes a deflection of some sort in the structure of the weigher. An opposing force is then applied to bring the structure back to its unloaded, datum, position. In the steady state, the applied force matches the opposing force and the latter can be measured by a secondary transducer.

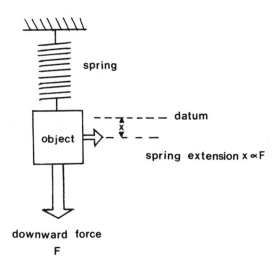

spring

datum

object

spring extension x ∝F

downward force

F

Fig. 6.2   Strain weigher (elastic weigher).

The earliest, and simplest, weighers are based on the arm balance (or steelyard) and these use the null balance principle. The unknown weight is balanced by a known weight, either by using a set of standard weights or by moving a known weight along the arm until a balance position is found. A more modern equivalent is shown in fig. 6.1b. The load is balanced on a hydraulic ram, whose position is monitored by a LVDT (see section 4.5.2). The position measurement is used to admit, or release, fluid until the null position is reached. The hydraulic pressure then indicates the load as the upwards force is (pressure × ram area). The hydraulic pressure can be measured by any of the transducers described in chapter 3.

The second technique is the strain weigher, shown in fig. 6.2. The applied load again causes a deflection of the weigher structure. This deflection is measured by a secondary transducer, and the load is deduced from a knowledge of the mechanical properties of the structure.

The simplest weigher is the spring balance. A spring is defined by its 'spring constant' which relates its extension to the load (e.g. 2 cm kg^{-1}). An applied load can easily be calculated by observing the spring extension and applying the spring constant. The simple spring balance can be made to give an electrical output by the use of a position measuring device, but in most industrial applications the need to attach peripheral equipment (such as piping) precludes large deflections. Most industrial weighers, however, are based on the strain weigher principle, but the secondary 'position' transducers are chosen to give minimal deflection.

6.1.2. Stress and strain

When any object is subjected to a force, deformation will occur. To be of any use, the relationship between the force and the deformation must be quantified. In fig. 6.3a a tensile force is applied to a rod of cross-sectional area A and length l. This produces a deformation in length Δl. The effect of the force is assumed uniform over the cross section of the rod, and intuitively the larger the area the force acts over, the less deformation will occur. The effect of the force is called the *stress*, and is defined as force per unit area. i.e.

$$\text{Stress} = F/A \tag{6.1}$$

Stress has the units of N/m^2 (i.e. it has the same units as pressure).

Intuitively, again, it would be expected that the change in length, Δl, will be proportional to the unstrained length l as the stress is equal at all points of the rod. The deformation is called the *strain* and is defined as the fractional change in length:

$$\text{Strain} = \Delta l/l \tag{6.2}$$

As Δl and l both have the dimensions of length, strain is a dimensionless quality. Often, however, the dimensionless unit 'microstrain' is used. A 10 m rod, for example, exhibiting a stress-induced length change of 4.5×10^{-5} m, is exhibiting 45 µstrain. Changes in dimensions are usually small, and the µstrain is a convenient unit.

In Fig. 6.3b a compressive force is applied to a rod, and this produces a reduction in length (assuming that the rod stays

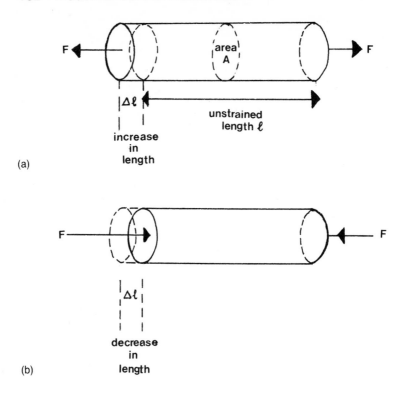

Fig. 6.3 Tensile and compressive strain. (a) Tensile strain. (b) Compressive strain.

straight, and bowing does not occur). Stress and strain are defined as above, with a change of sign for *F* and Δ*l* to indicate the change of direction.

If an object is subjected to an increasing stress, the strain will obviously increase, resulting in a relationship as indicated in fig. 6.4. In the region AB, the object behaves as a spring; the relationship is linear and there is no hysteresis (i.e. the object returns to its original size when the stress is removed; the deformation is not permanent). Beyond B, the object suffers permanent deformation. (In a bar or rod, necking starts to occur causing the stress to increase in a length of the object: see fig. 6.4b. The strain increases non-linearly with stress until it breaks at point C.)

The linear portion AB is called the elastic region and point B the elastic limit. Typically AB will correspond to 10 000 μ strain. Obviously all mechanical and civil structures must be kept within

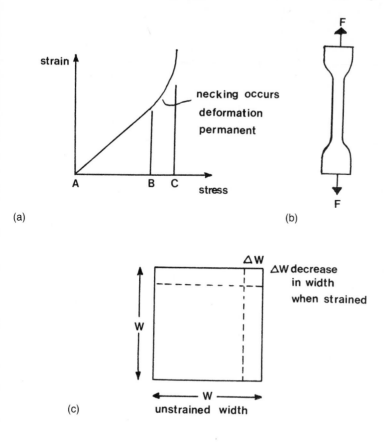

Fig. 6.4 The effect of stress on an object. (a) Relationship between stress and strain. (b) Necking in a rod under tensile stress. (c) Poisson's ratio; change in cross-sectional area with stress.

the elastic region.

The inverse slope of the line AB is called the elastic modulus or modulus of elasticity, and is constant for a specific material. For simple tensile or compressive stress the term Young's modulus is also used. The slope is defined by:

$$\text{Young's modulus} = \frac{\text{Stress}}{\text{Strain}} = \frac{F/A}{\Delta l/l} \qquad (6.3)$$

As strain is dimensionless, Young's modulus has the dimensions of $N\ m^{-2}$, i.e. pressure, and is commonly given in pascals.

Typical values are:

Steel	210 GPa
Copper	120 GPa
Aluminium	70 GPa
Plastics	30 GPa

When an object suffers tensile strain, not only does it experience a length change, but it also experiences a decrease in cross-sectional area. Similarly an object undergoing compressive strain will exhibit an increase in cross-sectional area. This effect is quantified by Poisson's ratio, usually denoted by the Greek letter ν. If an object has a length L and width W in its unstrained state, and experiences a change ΔL and ΔW when strained, the Poisson's ratio is defined as:

$$\nu = \frac{\Delta W/W}{\Delta L/L} \tag{6.4}$$

Typically ν is between 0.2 and 0.4.

Poisson's ratio can be used to calculate the change in cross-sectional area. In fig. 6.4, the unstrained area is $W.W$, and the change in area $2.W.\Delta W$ (neglecting the term ΔW^2). The fractional change is $2.\Delta W/W$, where ΔW can be calculated, knowing the tensile (or compressive) stress, via Young's modulus and Poisson's ratio.

6.1.3.   Shear strain

A third type of strain is shown in fig. 6.5. A force is applied to the top of a cube, causing it to distort. This is known as shear strain. To quantify shear strain we must first define shear stress.

Stress has the dimensions of force per unit area, so the shear stress on fig. 6.5 is defined as:

$$\text{Shear stress} = F/A \tag{6.5}$$

where A is the cross-sectional area of the face.

Shear strain is the ratio between the displacement, BC, and the height of the cube AB, i.e.

$$\text{Shear strain} = \frac{BC}{AB} = \tan\theta \tag{6.6}$$

As strains are small, shear strain approximates closely to θ radians.

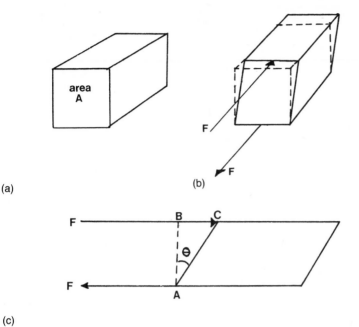

Fig. 6.5 Shear strain. (a) Unstrained object. (b) Object with shear strain. (c) Side view and definition of shear strain.

We can again define a modulus of rigidity, which has the same form as equation 6.3 except shear stress and shear strain are used. Typical values are:

Steel	75 GPa
Copper	45 GPa
Aluminium	25 GPa

6.2. Strain gauges

6.2.1. Introduction

The electrical resistance of a conductor such as that of fig. 6.6 is proportional to its length, and inversely proportional to the cross-sectional area, i.e.

$$R = \varrho \frac{L}{H.W} = \varrho \frac{L}{A} \tag{6.7}$$

where ϱ is a constant called the resistivity of the material.

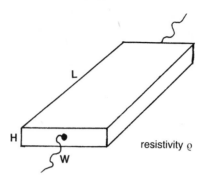

Fig. 6.6 Electrical resistance of a slab conductor.

When the conductor suffers tensile stress, L will increase (as determined by equation 6.3) and A will decrease (ΔW and ΔH being determined by equation 6.4). The resistivity also increases slightly as the material is deformed. The cumulative effect of these changes is to incease the resistance

$$R + \Delta R = (\varrho + \Delta \varrho) \frac{(L + \Delta L)}{(A - \Delta A)} \qquad (6.8)$$

Similar arguments show that a compressive stress will cause the resistance to decrease.

Ignoring second-order effects, the fractional change in resistance $\Delta R/R$ is proportional to $\Delta L/L$, i.e.

$$\frac{\Delta R}{R} = G \frac{\Delta L}{L} \qquad (6.9)$$

where G is a constant called the gauge factor.

Strain, however, is given by $\Delta L/L$, so equation 6.9 can be rewritten:

$$\Delta R = G.R.e \qquad (6.10)$$

where e is the strain exhibited by the conductor. This is the fundamental strain gauge equation.

Equation 6.10 allows us to relate an observed change in resistance to a strain (and hence to an applied force). A transducer using this principle is called a strain gauge. In practice, the force is not applied directly to the gauge. Usually the gauge is attached to some stressed member with epoxy resin. The strain experienced by the gauge will then be identical to the strain induced in the member.

6.2.2. Foil gauges

Practical strain gauges are not a slab of material as implied by fig. 6.6. They must, of necessity, be flimsy devices so that their strain matches exactly that of the object to which they are attached.

Strain gauges must also ignore strains in unwanted directions. An active axis is defined for any gauge, and this is the direction in which the sensitivity is highest. Similarly, a passive axis is defined along which the gauge is least sensitive (usually at 90° to the active axis). These are related by the cross sensitivity:

$$\text{Cross sensitivity} = \frac{\text{Sensitivity along passive axis}}{\text{Sensitivity along active axis}} \qquad (6.11)$$

In a well-designed strain gauge, cross sensitivities as low as 0.002 can be obtained.

Early gauges were constructed from thin wire, but modern gauges are photo etched from a metal film deposited on a thin polyester or plastic backing. This allows a device to be manufactured which has predictable properties. The backing material assists handling and attachment, and serves to insulate the gauge from the (probably metallic) object under test. These devices are called foil strain gauges.

A typical device is shown enlarged in fig. 6.7. This has a length of 8 mm, an unstrained resistance of 120 ohm and a gauge factor of 2. Standard gauge resistance values are 120 and 350 ohm; most have a gauge factor of about 2. Leads are attached to pads on the foil.

Gauges are available in lengths from about 0.25 mm up to about 50 mm, although 5–10 mm gauges are commonest. The gauge length should be chosen such that the strain is even along the gauge, although there are obvious handling problems with small devices. Small gauges also tend to have degraded performance.

Standard gauges can be used up to about 10 000 µstrain, although permanent resistance changes can occur with cyclical strains above about 2000 µstrain. Strains up to 10% (100 000 µstrain) can occur in plastic and can be measured by wire gauges constructed from annealed constantan.

Foil gauges are effectively glued to the object under test, but

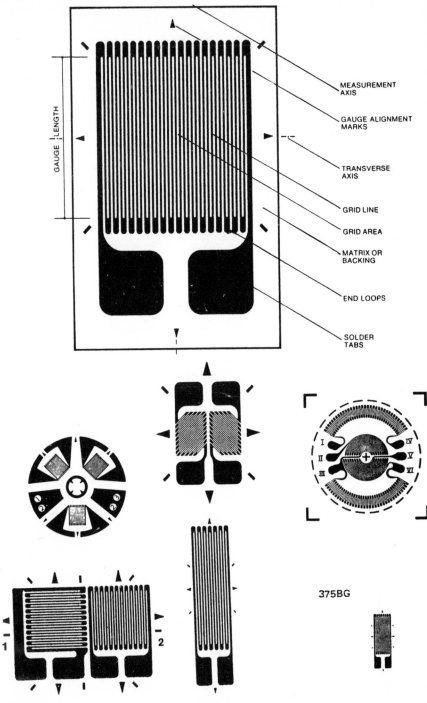

MEASUREMENT AXIS

GAUGE ALIGNMENT MARKS

TRANSVERSE AXIS

GRID LINE

GRID AREA

MATRIX OR BACKING

END LOOPS

SOLDER TABS

GAUGE LENGTH

375BG

Fig. 6.7 Typical strain gauges and terminology. Insert 375BG is shown full size. (Courtesy of Welwyn Strain Measurement Ltd, UK representatives of Measurement Group, Vishay, USA.)

the operation requires far more care than this simple statement might imply. It is imperative that the strain experienced by the gauge matches exactly the strain of the object. Many glues and epoxy resins have a 'creeping' property that allows strain relief in the gauge and causes responses as shown in fig. 6.8. This property is common in adhesives designed to give a flexible joint. Resins designed specifically for strain gauges should be used.

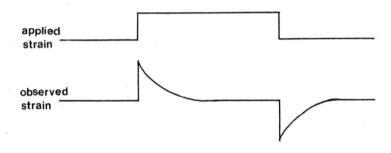

Fig. 6.8   The effect of 'creep' in the mounting adhesive.

The surface of the object must be clean and grease-free, and solvent degreasing is required just before attaching the gauge. It is essential to avoid touching the area with fingers and dragging in dirt from surrounding areas. Cleanliness is essential: any contaminated area can lead to a loss of strain coupling. The area next needs abrading with silicon carbide paper to give a good key for the resin.

Guide lines are now usually marked with a 4H pencil to assist with aligning the gauge. This is necessary as placing the gauge is a once-and-only operation; it must be right first time and cannot be skewed or slid into position. A conditioner is now applied to prepare the surface and remove any excess material from the guide lines.

Finally, the resin is applied in a thin layer with no entrapped air bubbles. The gauge must not be contaminated by finger contact and should be peeled down in the correct position first time, immediately after the application of the resin. Gauge handling is simplified by attaching a short length of adhesive tape to the top (non-contact) face of the gauge. The tape should not be removed until the resin has set. If weatherproofing is required, a coating of silicon rubber compound can be applied after lead connections have been made.

The change in resistance in a strain gauge is very small. Strain values in metals are normally less than 1000 μstrain (failure can occur at 10 000 μstrain). For a typical gauge of 120 ohm resistance and gauge factor 2, from equation 6.10 this corresponds to a resistance change of 0.24 ohm. In many applications resistance changes as low as 0.01 ohm must be measured.

Resistance also changes with temperature (see section 2.4). These temperature effects are similar to, or larger than, the strain-induced effects. Bridge circuits, described in section 6.3, provide temperature compensation. Temperature changes will also cause dimensional changes in the object under test, which are indistinguishable from strain-induced changes. Temperature-compensated gauges are available with coefficients of linear expansion matched to those of common materials such as mild steel.

6.2.3. Semiconductor strain gauges

The change in resistance with stress (as opposed to the change in resistance with strain discussed so far) is called the piezo-resistance effect. The effect is small in metals, but is significant in suitably doped semiconductor materials. Devices based on this principle are called semiconductor strain gauges (the external strain inducing a stress in the crystal).

Their main advantages are large gauge factors, typically 250, and small size (less than 0.5 mm), which allow them to be used, for example, in medical transducers. They are, however, very temperature sensitive. Both the gauge factor and the actual gauge resistance vary with temperature. The gauge factor also varies with the strain in the crystals. Semiconductor gauges are therefore usually restricted to applications requiring measurement of small, localised, dynamic strains.

6.3. Bridge circuits

6.3.1. Wheatstone bridge

The small change in resistance in a strain gauge is superimposed on a much larger unstrained resistance. Typically the change will

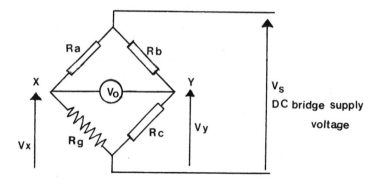

Fig. 6.9   Measurement of strain with a Wheatstone bridge.

represent one part in 1000 to 5000, which makes it difficult to produce an output voltage (or current) by direct means.

The classical laboratory method of measuring unknown resistance is the Wheatstone bridge shown in fig. 6.9. In the normal laboratory method, R_a and R_b are equal, and the calibrated resistance box R_c is adjusted until V_o is zero (as measured by a sensitive galvanometer). Voltages V_x and V_y must be equal, so:

$$\frac{R_g}{R_g + R} V_s = \frac{R_c}{R_c + R} V_s \qquad (6.12)$$

or

$$R_g = R_c \qquad (6.13)$$

the value of the unknown resistor can be read from the calibrated resistance box R_c.

In strain-gauge work, however, we are not so much interested in the actual *value* as in its *change* from some norm. Suppose R_b and R_c are made fixed, and equal, resistances, such that V_y is half V_s. If R_a is made equal to the unstrained resistance of the gauge, V_x will also be 0.5 V_s, and V_o will be zero. If the gauge is strained, R_g will increase to $R_g + \Delta R$ causing a rise in V_x, and a non-zero reading of V_o. We must relate this change in V_o to the strain, e, causing ΔR. Let us assume that the unstrained gauge resistance is R, and to simplify the calculation we will represent the fractional change in resistance ($\Delta R/R$) by x. Assuming the device measuring V_o draws minimal current:

$$V_y = 0.5V_s \tag{6.14}$$

$$V_x = \frac{R_g}{R_g + R_a} V_s \tag{6.15}$$

$$= \frac{R(1 + x)}{R(1 + x) + R} V_s \tag{6.16}$$

$$= \frac{(1 + x)}{(2 + x)} V_s \tag{6.17}$$

Now $V_o = V_x - V_y$ $\tag{6.18}$

$$= \frac{(1 + x)}{(2 + x)} V_s - \frac{V_s}{2} \tag{6.19}$$

$$= \frac{V_s x}{2(2 + x)} \tag{6.20}$$

Typically, x will be negligible compared with 2 (for the very large strain of 1000 μstrain calculated in section 6.2.2, x was 0.002). Equation 6.20 thus approximates to:

$$V_o = \frac{V_s x}{4} = \frac{\Delta R}{4R} x \tag{6.21}$$

But $\Delta R = eGR$, so:

$$V_o = \frac{eGR_x}{4R} V_s \tag{6.22}$$

$$= \frac{eG}{4} V_s \tag{6.23}$$

Equation 6.23 implies that the change in output voltage is linear and proportional to the strain, and the sensitivity is dependent on the supply voltage V_s and the gauge factor G. It is independent of the unstrained gauge resistance. This result is, however, conditional on the assumption that x is small (which is usually a more than reasonable assumption).

It is instructive to put some values into equation 6.23. For the gauge of fig. 6.2, G has the value 2. Choice of bridge voltage will be discussed later, but let us assume $V_s = 24$ V. For 1000 μstrain, we get:

$$V_o = \frac{1000 \times 10^{-6} \times 2 \times 24}{4}$$

$$= 12 \text{ mV}$$

i.e. very small. The bridge output voltage requires considerable amplification before it can be used for indication or control.

6.3.2. Temperature compensation

The circuit of fig. 6.9 cannot distinguish between resistance changes caused by strain and those caused by temperature effects. In the arrangement of fig. 6.10a, gauge 1 is aligned with the strain, and gauge 2 is aligned across the strain. Gauge 1 will exhibit a change in resistance as predicted by equation 6.10 (plus temperature effects) whereas gauge 2 will only have a small change due to strain (as predicted by equation 6.11) plus a change

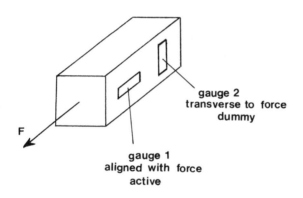

(a)

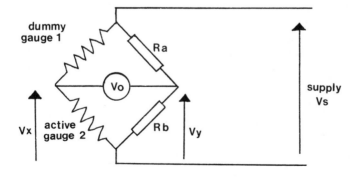

(b)

Fig. 6.10   The use of a dummy gauge for temperature compensation. (a) Mounting arrangement. (b) Wheatstone bridge circuit.

due to temperature. As both gauges are in close proximity it is reasonable to assume that both are at the same temperature.

Let x be the fractional change in resistance due to strain (occurring in gauge 1 only) and t the fractional change in resistance due to temperature. As before, $V_x = 0.5\ V_s$:

$$V_x = \frac{R_1}{R_1 + R_2} V_s \tag{6.24}$$

$$= \frac{R(1 + x)(1 + t)}{R(1 + x)(1 + t) + R(1 + t)} V_s \tag{6.25}$$

Terms in $(1 + t)$ cancel, leaving:

$$V_x = \frac{R(1 + x)}{R(1 + x) + R} V_s \tag{6.26}$$

$$= \frac{(1 + x)}{(2 + x)} V_s \tag{6.27}$$

which is the same result as equation 6.20, leading, therefore, to the same result as equation 6.23. The inclusion of the dummy gauge has compensated completely for resistance changes caused by variations in temperature.

6.3.3. Multigauge bridges

Many applications utilise gauges in all four bridge arms. In fig. 6.11a, R_1 and R_3 are aligned with the strain, and R_2 and R_4 transversely. Connected as shown in fig. 6.11b, V_x will increase and V_y decrease with strain, giving twice the sensitivity of fig. 6.10. R_3 and R_4 act as dummy gauges. As all four gauges are affected equally by temperature changes, the bridge is temperature compensated.

The arrangement of fig. 6.11c is used to detect bending strain and ignore axial strain. As the object bends, material above the neutral axis will suffer positive strain and that below negative strain. If the gauges are again connected as in fig. 6.11b, bending strain as shown will cause V_x to increase and V_y to decrease.

Axial strain affects all four gauges equally, and consequently will not affect V_o. Similarly, temperature changes will cause equal changes in all gauges and will be ignored.

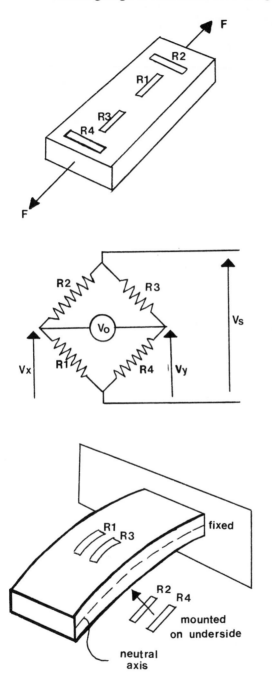

Fig. 6.11 Four-gauge applications. (a) Four-gauge mounting. (b) Four-gauge bridge giving increased sensitivity. (c) Measuring bending strain in presence of axial strain.

6.3.4.   Bridge balancing

Gauges are manufactured to a resistance tolerance of about 0.5%; about ±0.6 ohm for a 120-ohm gauge. As this is greater than the change caused by strain, some form of zeroing will be required so that V_o is zero in the unstrained state. It is possible to incorporate this in the amplifier stages after the bridge, but it is common to include bridge balancing in the bridge circuit itself.

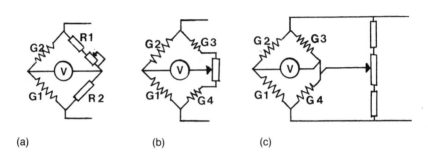

(a)   (b)   (c)

Fig. 6.12   Methods of bridge balancing. (a) One leg balancing (R1<R2). (b) Apex balancing. (c) Parallel balancing.

There are three common ways of achieving this, shown in fig. 6.12. In all three, the potentiometer is adjusted to zero V_o in the unstrained state. Arrangements 6.12b and 6.12c are preferred if the potentiometer is remote from the bridge, as errors induced by the resistance of the interconnecting cables (and changes in the cable resistance caused by temperature) will not cause a span or zero shift in V_o.

6.3.5.   Bridge connections

Equation 6.23 predicts that the bridge sensitivity is proportional to the bridge supply voltage. Increasing the supply voltage, however, increases the bridge dissipation (which rises as the square of the voltage). In the extreme this could permanently damage the gauges, but errors caused by gauge heating will occur at much lower powers.

Gauge manufacturers specify a maximum dissipation, gauge current or bridge voltage. In general, higher resistance gauges can be used at higher voltages. Typical bridge voltages and currents are 15–30 V and 20–100 mA.

Equation 6.23 also indicates that V_s must be stabilised. If the bridge is remote from the power supply unit (PSU), bridge-balancing potentiometer and amplifer electronics, error can be introduced from the resistance of the connecting cables.

There are many possible ways of compensating for lead resistance, and one solution is shown in fig. 6.13. The bridge supply is provided on lines 1 and 2 from a stabilised power supply. The actual bridge voltage is sensed on lines 3 and 4, which provide feedback so the PSU gives the correct voltage at the bridge (rather than at the PSU terminals). Leads 5 and 6 are part of the apex balancing circuit, but as both leads have the same resistance, no error is introduced. The other apex of the bridge is brought back on lead 7. Leads 3, 4 and 7 are sensing leads carrying minimal current, so the errors caused by the lead resistance are negligible.

6.3.6. Bridge amplifiers

The bridge output voltage of a few millivolts needs to be amplified before it can be used. Care must be taken to avoid common mode induced noise, and it is usual to employ a differential amplifier similar to fig. 6.14. This circuit is discussed

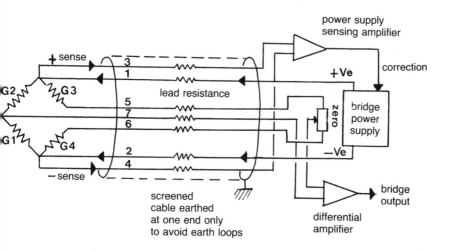

Fig. 6.13   Cabling of remote bridge.

in detail in chapter 1, Volume 2, but if $R_a = R_b = R_1$, and $R_c = R_d = R_2$, the output voltage is:

$$V_o = \frac{R_2}{R_1} (V_2 - V_1) \tag{6.28}$$

$$= -\frac{R_2}{R_1} V_b \tag{6.29}$$

where V_b is the bridge output voltage. The circuit amplifies the difference between its input terminals and ignores common mode voltage. The gain is simply $- R_2/R_1$.

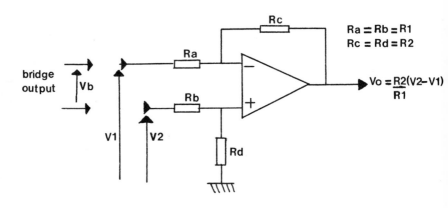

Fig. 6.14 Differential amplifier.

Any real-life voltage source can be represented by a perfect voltage source in series with a resistor (Thévenin's theorem). The voltage from each apex of a bridge such as fig. 6.15a can be considered as an open circuit voltage V_x in series with a resistor. The value of this resistor is simply $R_g/2$ where R_g is the nominal strain-gauge resistance.

In fig. 6.15c, the bridge is connected direct to the amplifier without resistors R_a and R_b of fig. 6.14. Thévenin's theorem allows us to draw this circuit as fig. 6.15d with the equivalent circuit for each bridge apex voltage. The output voltage from the amplifier is:

$$V_o = -\frac{2R}{R_g} (V_y - V_x) \tag{6.30}$$

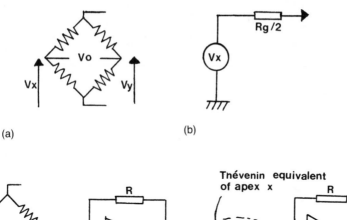

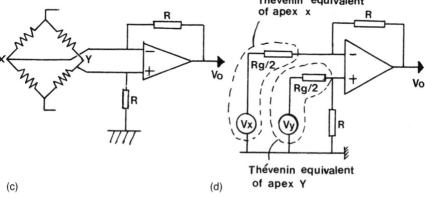

Fig. 6.15 Thévenin representation of Wheatstone bridge. (a) Bridge circuit. (b) Thévenin equivalent circuit of an apex. (c) Bridge connected to differential amplifier. (d) Equivalent circuit.

6.3.7. Torque measurement

Torque is a twisting force, usually encountered on shafts, bars, pulleys and similar rotational devices. It is defined as the product of the force and the radius over which it acts. The shaft in fig. 6.16a, for example, is experiencing a torque of 4.2 Nm.

Torque produces distortion of a shaft which can be visualised as a 'wind-up' along the shaft. This wind-up is a form of shear strain, and in fig. 6.16a the shaft has been distorted by an angle θ. Torque-induced shear strain is defined by θ if θ is expressed in radians.

A shear modulus G can be defined for the shaft. T and θ are then related by the expression:

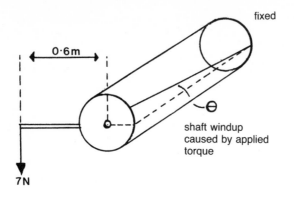

(a)

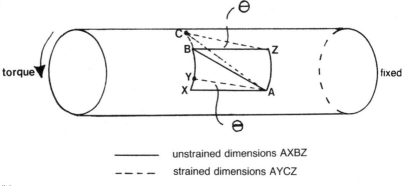

(b)

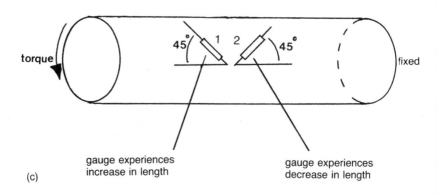

(c)

Fig. 6.16 Measurement of torque. (a) Definition of torque and torque-induced strain. (b) Dimensional effects of torque-induced strain. (c) Mounting strain gauges to measure torque-induced strain.

$$T = \frac{\pi G r^3}{2} \theta \tag{6.31}$$

where r is the shaft radius. Note that the shear strain varies inversely as the third power of the shaft radius.

Shear strain induced by torque can be measured by a strain gauge mounted at 45° to the axis as in fig. 6.16. In the unstressed state, the gauge will lie along the line AB. When torque is experienced, the gauge lies along the line AC. The line AC is longer than AB, so the gauge experiences a strain:

$$e = \frac{AC - AB}{AB} \tag{6.32}$$

$$= \theta/2 \tag{6.33}$$

where θ is small (and expressed in radians).

In practice, two or more gauges are used to provide increased sensitivity and temperature compensation. In fig. 6.16c, for example, gauge 1 is experiencing tensile strain and gauge 2 compressive strain.

There are practical problems in connecting measuring equipment to gauges on a rotating shaft. This is usually achieved by slip rings or by using an AC bridge supply and coupling both bridge supply and output signal via a transformer. If the centrifugal forces are not large and an out-of-balance load can be tolerated on the shaft, an entire battery-based power supply, amplifier and low-powered radio telemetry transmitter can be strapped to the shaft.

Torque measuring devices are used for determining the power of engines, motors, and other rotating devices. Such instruments are commonly called dynamometers.

6.4. Magnetoelastic devices

The strain gauge is not the only force measuring device. An important class of devices uses a change in magnetic permeability with applied force as a load sensing mechanism. The principle, known as the magnetoelastic effect, is shown in fig. 6.17.

In fig. 6.17a, a magnetic field, H, is applied at 45° to a slab of steel. This induces equal flux densities B_v and B_h, giving a net vector $\mathbf{B}$ in the same direction as H. Permeability, however, is decreased by an applied force. In fig. 6.17b the applied force

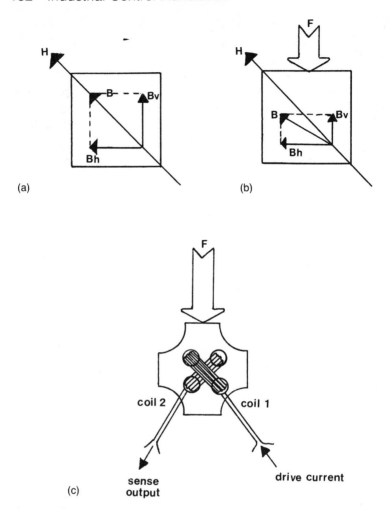

Fig. 6.17 Magnetoelastic force transducer. (a) Unstrained state, B aligns with H. (b) Strained state, B no longer aligns with H. (c) Practical device.

causes a reduction in B_v (B_h being unchanged). The net flux vector **B** is no longer aligned with H, and the deviation θ is related to the force F.

This principle can be applied as shown in fig. 6.17c. Two coils are wound through holes in a transformer steel core. The coils are precisely aligned at 90° so in the unstressed state there is no coupling between them. Coil 1 is driven by an AC source, but no voltage is induced in coil 2.

When a force is applied, the shift of the **B** vector causes a voltage to be induced in coil 2; the magnitude of the voltage being dependent on the applied force.

Magnetoelastic transducers are primarily marketed by the Swedish company ASEA under the trade name Pressductor. These devices are physically much more robust than strain gauge devices and have an output signal of several hundred millivolts (compared with the few millivolts of a strain gauge bridge) and as such are less affected by noise. The drive electronics are, however, more complex, and the device's non-linear response requires more complex linearisation circuits.

6.5. Loadcells

Strain gauges and (to a lesser extent) magnetoelastic devices are strain measuring devices. A weighing device is required to measure weight (or more pedantically force). A force measuring device is called a loadcell, and usually works by converting an applied force to a measurable strain. Common sensing arrangements are shown in fig. 6.18. These would all be enclosed in a protective case. All utilise four strain gauges for maximum sensitivity and temperature compensation (magnetoelastic transducers all use a sensor similar to that shown in fig. 6.17c).

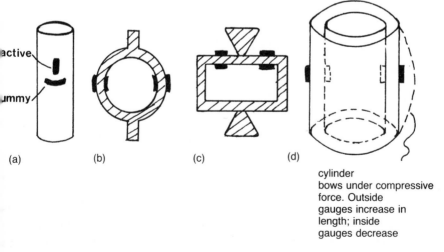

cylinder
bows under compressive
force. Outside
gauges increase in
length; inside
gauges decrease

Fig. 6.18 Various load sensors with gauge mounting positions. (a) Simple column. (b) Proof ring. (c) Proving frame. (d) Cylinder.

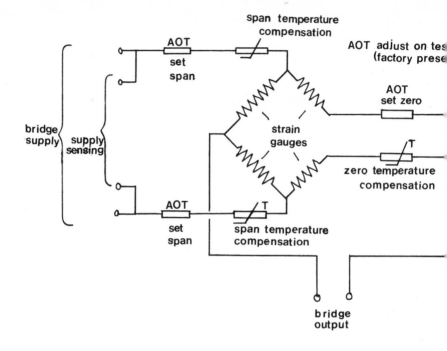

Fig. 6.19 Schematic diagram of industrial weighing loadcell. Note that a six-core screened cable is required.

The column type is only used for large loads as the strain is small. The proof ring, proving ring and cylinder give reasonable outputs at low loads. All rely on two strain gauges in compression and two in tension. The gauges are arranged in a circuit, as shown in fig. 6.19. The span resistors are adjusted on test to give a predetermined output for the rated load. The temperature compensation resistors compensate for changes in Young's modulus of the cell with temperature (not for bridge changes which are, of course inherently ignored by the bridge).

Coupling of the load to the cell requires care. A typical arrangement is shown in fig. 6.20. A pressure plate, effectively a metal/rubber/metal sandwich, applies the load to a knuckle on the top of the loadcell and avoids errors caused by slight misalignment. A flexible diaphragm seals the cell against dust and the weather. Practical devices are shown in fig. 6.21. The cell in fig. 6.21a is a compressive cell and that in fig. 6.21b is a tensile cell.

Only rarely is a single loadcell used as it is difficult to keep a

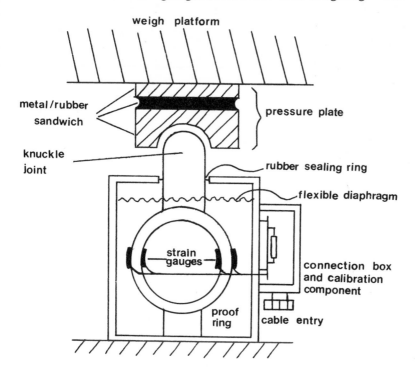

Fig. 6.20 Construction of typical loadcell.

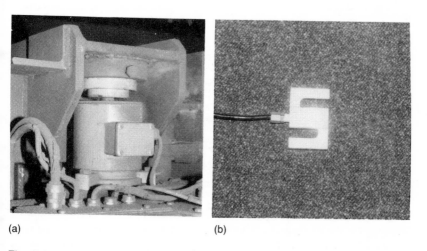

(a)   (b)

Fig. 6.21 (a) Practical load cell. Note knuckle joint & pressure plate. (b) S type load cell.

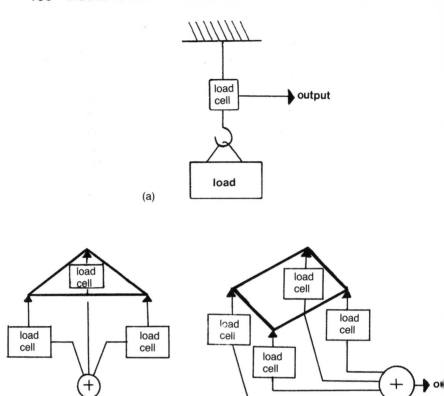

Fig. 6.22 Mechanical coupling of loadcells. (a) Single-point support. (b) Three-point support (preferred). (c) Four-point support. Care must be taken to ensure cells are always loaded.

load vertical on a single point support. Guide rods and tie bars can be used, but care must be taken to avoid errors from friction. Single cells can be used in suspension weighers (fig. 6.22a). Multicell systems are commoner, the best arrangement being the three-cell system of fig. 6.22b. This inherently spreads the load roughly equally among the three cells, whose outputs are summed electronically. A four-cell system such as in fig. 6.22c can be used, but care must be taken to ensure that the load is shared at all times among the cells and that no cell ever loses contact with the load platform.

The choice of a loadcell is obviously determined by the maximum expected load, but there are other factors to consider. An overload capacity should be included to cover mechanical failures and shock loads from, say, material falling into a hopper resting on loadcells. Typical overload allowance is 100–500%. The system should also be examined for possible side loads, which can cause damage to the sealing diaphragms and the sensing element itself. Acceleration of a moving weigh car can place severe side loads on loadcells supporting a heavy weight of material.

Usually the loadcells are the only electrical route to ground from the weigh platform. It is advisable to incorporate a flexible earth strap to the platform, not only for electrical safety but also to provide a route to ground for welding currents should mechanical repairs be necessary on the weigh platform. Welding current passing through a loadcell body will cause instant failure.

6.6. Weight controllers

A weighing system is usually more than a collection of loadcells and a display system. Figure 6.23 illustrates two common weighing systems. Figure 6.23a is a Taring system, used where a 'recipe' of several materials is to be combined in a hopper. The gross display shows the total weight indicated by the cells, including the weight of the hopper itself.

When the tare button is pressed, the current gross weight is stored and subtracted from the gross to give a net weight display. This can be used to obtain a weight for each new material in a batch, the tare button being pressed before every addition.

Figure 6.23b is a batch feeder system, where material is fed to a hopper by a variable speed vibrating feeder. The speed is controlled by the weigh system along the lines of fig. 6.23c. A two-speed system is commonly used, with a changeover from fast to dribble at some fixed point before the target weight. The feeder turns off at a second fixed point just before the target weight to allow for material in flight between the feeder and the hopper. This is known variously as anticipation, preact or in-flight compensation. Setting up of the four parameters of fig. 6.23c is an obvious compromise between speed and accuracy.

Other features often found in weight control schemes are indication of stable weight (sometimes called motion detection),

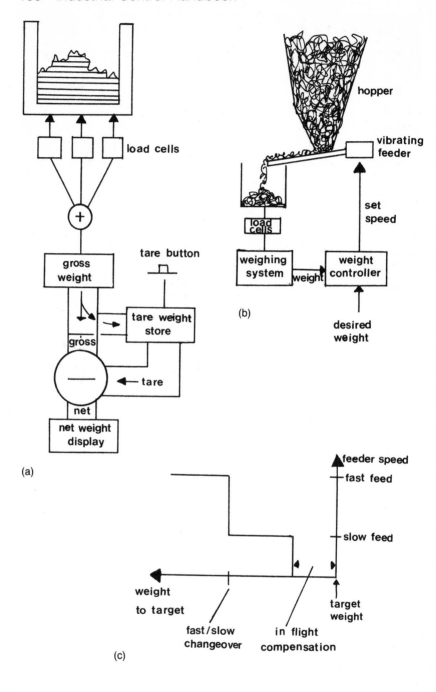

Fig. 6.23 Weighing control systems. (a) Tare weigher. (b) Batch feeder system. (c) Feeder speed set-up. (d) Photograph of batch feeder.

Fig. 6.23 contd.

automatic taring and storage of batch menus for automatic mixing.

6.7. Belt weighers

Solid material is often carried by conveyor belts, and an indication is usually required of feed rate in weight per unit time (e.g. tons per minute). This can be provided by the scheme of fig. 6.24.

The conveyor passes over a weigh platform of known length L metres, and the conveyor speed is measured by a suitable device (e.g. a tachogenerator). If the weight controller indicates a load of M kg, and the belt is moving at V m s^{-1}, the feed rate is simply $M.V/L$ kg s^{-1}.

The feed rate can obviously be used to control either the feeder or the belt speed, but the inherent transit delay can make it a difficult system to control.

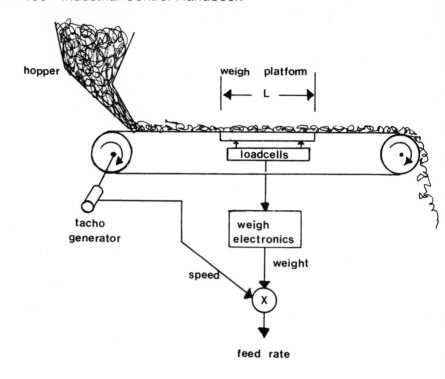

Fig. 6.24   Belt weigher schematic.

Chapter 7
Level measurement

7.1. Introduction

Wherever liquids, or bulk solids, are used, stored or conveyed, some type of level indication will be needed. There are many techniques for the measurement of level, and the choice of sensor is more difficult than for other process variables.

Level is often used to infer volume. This is straightforward with a container of uniform cross section such as the tank of fig. 7.1a where the volume is given by:

$$V = hA \tag{7.1}$$

The relationship for the pressure vessel of fig. 7.1b is, however distinctly non-linear, as shown in fig. 7.1c.

Level can also be used to infer mass by calculating the volume

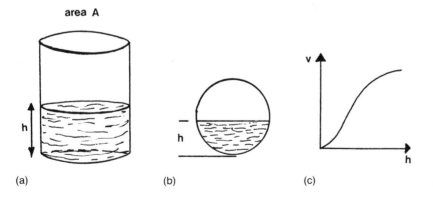

Fig. 7.1 Relationship between level and volume. (a) Cylindrical tank with linear relationship between depth and volume. (b) Spherical pressure tank. (c) Relationship between depth and volume for pressure tank.

and multiplying by the density. This is subject to the same problems of container geometry as volume measurement, plus the additional complication that the density may be affected by temperature or the chemical composition of the liquid/bulk solid concerned.

In many applications where volume or mass measurement is really needed, the simplest solution is often simply to weigh the tank, container or silo by suspending it from loadcells or mounting it on a weigh scale. Care must be taken to avoid friction effects from inflow or outflow pipe connections, but flexible couplings are available similar to those on the powder silo of fig. 7.2. Weighing gives volume or mass measurement which is unaffected by the shape of the container.

The conditions inside the tank must also be considered. Temperature, pressure and corrosive atmospheres tend to affect level sensors more than, say, resistance thermometers or thermocouples. The liquid (or solid) characteristics must also be

Fig. 7.2 Level measurement by weight measurement. Note flexible hoses on pipes.

considered. A fluid that wets or clings to a sensor will tend to give a possibly dangerous high indication. Level measurement will also be adversely affected by turbulence or frothing on the surface (a domestic example of which is self-perpetuating 'hammer' in hot-water header tanks where ripples affect the float-operated shut-off valve). It may be necessary to provide a protected 'harbour' or still well for the level sensor.

A tank containing liquid can also resonate. A tank of diameter L metres will oscillate with period:

$$T = 2\pi\sqrt{\frac{L}{2g}} \quad \text{(secs)} \tag{7.2}$$

Note that this is independent of density. Rectangular vessels can exhibit two differing resonant periods. Hydraulic resonance is often observed by parents of young children at bath time! Inserting the length of a bath (1.5 m) into equation 7.2 gives a period of about 1.7 s.

There are, in general, four interfaces that need to be considered for level control:

(a) liquid/gas;
(b) solid/gas;
(c) liquid 1/liquid 2 (non-miscible, e.g. water/oil);
(d) solid/liquid (rare, e.g. identifying sludge depth in oil tank).

Most applications cover the first two types. Level sensors, in general, work by identifying the position of the interface or responding in some way to the bulk of one of the materials.

7.2. Float-based systems

Float-based measurement systems are the simplest level transducers. The basic principle was discovered many centuries ago by Archimedes, and is illustrated in fig. 7.3a. A floating body experiences two forces: a downward force of gravity and an opposing force due to its buoyancy.

$$\text{Downward force} = g \times \text{mass of float} \tag{7.3}$$

$$\text{Upward force} = g \times \text{mass of liquid displaced} \tag{7.4}$$

In equilibrium these balance, so

$$g \times \text{mass of float} = g \times \text{mass of liquid displaced} \qquad (7.5)$$

$$\text{Mass of liquid displaced} = \varrho \times A \times d \qquad (7.6)$$

where ϱ is the liquid density, A the float area and d the

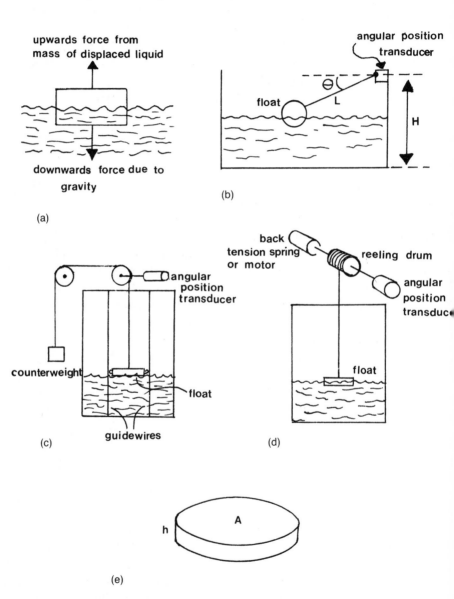

Fig. 7.3 Float–based level measurement. (a) Archimedes' principle. (b) Lever arm float-level sensor. (c) Counterweight float. (d) Reeling drum float. (e) Ideal float shape: large surface area, low height.

submerged depth of the float. The float is therefore submerged at a constant depth:

$$d = \frac{M}{\varrho A} \tag{7.7}$$

where M is the mass of the float.

Simple float systems are based on a rigid arm as shown in fig. 7.3b,c, and convert the liquid level to an angle which can be measured by an angular position transducer (commonly a potentiometer, see section 4.2). While these arrangements are simple, the output is non-linear with the liquid depth being given by:

$$h = H - L \sin \theta \tag{7.8}$$

The arrangements of fig. 7.3c,d use a vertically moving float and accordingly give a linear output. Figure 7.3c uses a counterweight to couple the float to the position transducer, and fig. 7.3d a torque-controlled electric motor (e.g. a stalled DC motor operating in current limit). Other possibilities include a spring-powered tape-up drum. In each case the output is a signal proportional to level.

Equation 7.7 shows that the float submersion depth is sensitive to changes in M and ϱ (A presumably remains constant). At low levels the masses of the floats in figs. 7.3c and d increase because of the length of the suspension wire. To minimise this error (and errors due to changes in ϱ), A should be large implying a 'cheesebox'-shape float as in fig. 7.3e. Errors can also occur if the suspension wire is not vertical due to, say, liquid flow in the tank. Restraining guide wires or a float running on a guide tube should be used in these circumstances.

Figure 7.3 requires either the position measuring device to be mounted inside the tank (with attendant maintenance problems, and totally impossible in applications where the tank atmosphere is corrosive or explosive) or the float displacement conveyed outside the tank via flexible bellows. Both cases present difficulties if the tank is substantially above atmospheric pressure.

Figure 7.4 shows a float-based system that can operate at very high pressures, albeit with a limited measurement range. A ferromagnetic float is contained in a glass tube similar to a sight glass (which is itself the simplest level gauge). The float moves within the coils of an LVDT position transducer (see section 4.5.2) to give an electrical output dependent on level.

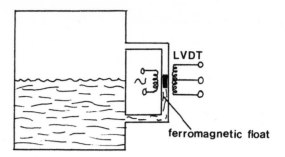

Fig. 7.4   Float level measurement usable at high pressures.

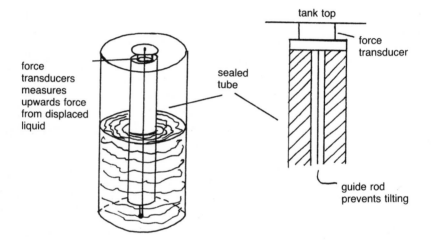

Fig. 7.5   Level measurement with force transducer.

Figure 7.5 is an alternative float transducer. In this application the 'float' is a fixed sealed tube connected to the top of the tank via a force transducer (e.g. a loadcell, see section 6.5). By equation 7.4 the float will experience an upward force dependent on the length of tube submerged. i.e. the liquid level. Having no moving parts, this arrangement is attractive in dirty locations. Care must again be taken to keep the tube vertical. The same technique can be used to twist a torque tube whose shear strain can be measured.

Surprisingly, floats can also be used to measure bulk solid level via the principle of fig. 7.6 (which also shows that 'level' is an imprecise term for bulk solids!). The technique does not provide

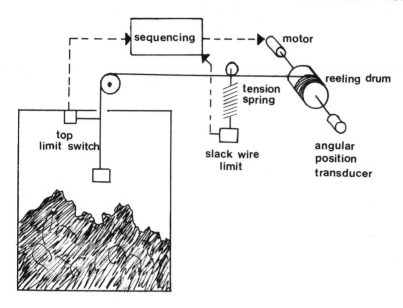

Fig. 7.6   Measurement of bulk solid level.

a continuous indication of level, but a sampled level at regular intervals. The 'float' is wound to the top of the silo, then lowered until the slack wire limit makes. The length of supporting cable paid out then gives the distance between the solid surface and the top of the silo. The depth of solid can be found by subtraction from the silo height.

7.3.   Pressure-operated transducers

7.3.1.   Direct measurement

A very common way of measuring level is the transformation of level to a differential pressure which can be converted to an electrical signal by any of the transducers in chapter 3. The basic principle is shown in fig. 7.7a. The absolute pressure at the bottom of the tank has two components: atmospheric pressure, and pressure due to the head of liquid. The absolute pressure is therefore given by:

$$P = \varrho gh + \text{atmospheric pressure} \tag{7.9}$$

where ϱ is the density of the liquid.

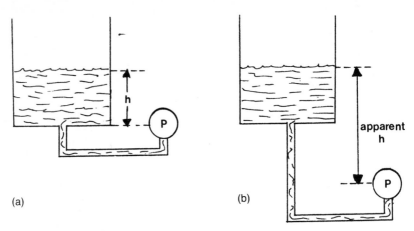

Fig. 7.7 Level measurement from hydrostatic pressure. (a) Gauge pressure proportional to level. (b) Head error.

In practice, a gauge pressure transducer is used which measures pressure with respect to atmosphere. The gauge pressure is given by:

$$P = \varrho g h \qquad (7.10)$$

which is linearly related to liquid level (and is independent of the tank shape and construction). One problem, however, is that the level, h, is measured with respect to the level of the pressure transducer itself and not the tank bottom. The system of fig. 7.7b will therefore indicate a level far in excess of its true value. Zero suppression can, of course, be used to give a true reading if the difference in height between the tank bottom and the pressure transducer is known.

Simple gauge pressure transducers cannot be used where the tank is at some pressure other than atmosphere. In these circumstances a differential transducer must be used with one leg connected to tank bottom (below the liquid surface) and one to the tank top as in fig. 7.8a. The pressure seen by the transducer is then linearly related to tank level because the static pressure on both legs cancels.

For fig. 7.8a to work, the LP leg must be kept totally free of liquid. This is often difficult to ensure, particularly in boiler applications where the space above the liquid is a condensable vapour. Figure 7.8a also retains the problem that a height differential between tank bottom and the transducer will give an

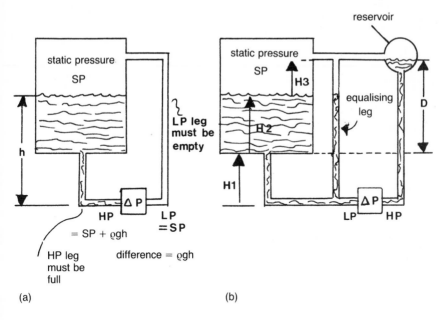

Fig. 7.8 Differential pressure-based level measurement in pressurised tanks. (a) Simple method. (b) Level measurement with condensable vapour.

offset in the output signal. Figure 7.8b overcomes both these problems by deliberately filling the leg to the tank top with liquid. In boiler level applications this can be arranged to occur naturally by keeping the measuring pipes unlagged and providing a small reservoir at the top of the tank top leg.

The pressure at the tank bottom leg is given by:

$$LP = \varrho_1 g(H_1 + H_2) + \varrho_2 g H_3 + SP \tag{7.11}$$

where ϱ_1 is the liquid density, ϱ_2 is the vapour density (often negligible), and SP is the static tank pressure.

The pressure in the tank top is given by:

$$HP = \varrho_1 g(H_1 + H_2 + H_3) + SP \tag{7.12}$$

The differential pressure is therefore:

$$DP = H_3(\varrho_1 - \varrho_2)g \tag{7.13}$$

If D is the height difference between the two tank tappings, this can be rearranged to:

$$DP = (D - H_2).(\varrho_1 - \varrho_2)g \tag{7.14}$$

which is independent of the transducer position H_1. In most applications the terms in ϱ_2 can be neglected.

Note that the LP leg of the transducer goes to the tank bottom and the transducer output is reversed, giving full scale for zero level and zero signal for maximum level. This can, of course, be reversed by suitable signal conditioning.

7.3.2. Gas reaction methods

The pressure-based methods in the previous section put some constraints on the user. In particular the pressure transducer must be at, or below, the lowest level in the tank. This is not always feasible, particularly with totally pneumatic systems. There can also be problems with applications where the liquid is corrosive or at a high temperature, as the fluid must come into direct contact with the pressure transmitter diaphragm. Level measurement in

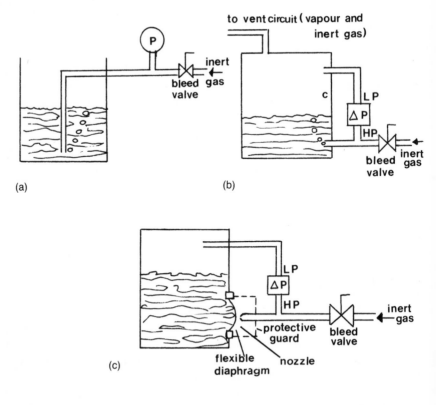

(a)

(b)

(c)

Fig. 7.9   Gas reaction methods. (a) Gas bubbler in open tank. (b) Sealed tank. (c) Sealed tank with flexible diaphragm.

the food industries must also avoid stagnant liquid in measuring lines as this can be an ideal breeding ground for bacteria.

These problems can be overcome by the techniques of fig. 7.9a. An inert gas is bubbled at a slow rate into the bottom of the tank. The pressure required to maintain flow is identical to the static pressure at the pipe entry and hence proportional to tank level. The pressure is measured by a conventional pressure transmitter, but because the measuring line only contains the purge gas, errors due to changes in level are negligible. The purge gas, usually argon or nitrogen, also keeps the measuring lines free of liquid, preventing corrosion or potential bacteria contamination.

Figure 7.9a is shown for a tank open to atmosphere with a gauge pressure transmitter. With a pressurised tank, a differential pressure transmitter is used, as in fig. 7.9b. Obviously the purge gas supply pressure must exceed the tank pressure, and some vent mechanism must be used to avoid raising the tank pressure with the purge gas.

A common variation on the principle is shown in fig. 7.9c. A measuring head with a flexible diaphragm is placed in the tank side, and controls the pipe pressure via a flapper/nozzle link. The purge gas pressure at the flapper/nozzle is then dependent on the diaphragm pressure and hence tank level. The gas vents to atmosphere and does not enter the tank. This arrangement can be used as shown for open tanks or modified for pressurised tanks by taking the LP leg of a differential pressure transmitter to the tank top. In practice, the differential pressure transmitters would be positioned to avoid the LP leg filling with condensed vapour (e.g. by mounting on the tank top).

7.3.3. Collapsing resistive tube

Figure 7.10 shows an interesting pressure-based level transducer that gives a direct electrical output. A flexible sealed tube is lined on opposite inside faces with resistive material. The tube is maintained at a pressure slightly above tank pressure. The tube inflates, keeping the resistive faces apart. When the tube is submerged, it collapses below the surface, bringing a length of the resistive faces together. The electrical resistance of the tube therefore varies with level. The device can, with care, also be used for bulk solid level measurement.

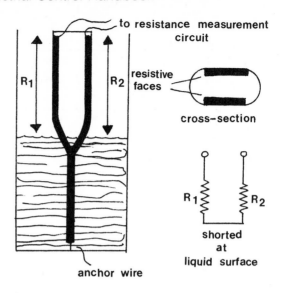

Fig. 7.10 Collapsing tube method.

7.4. Direct electrical probes

7.4.1. Capacitance sensing

The basis of a capacitance sensing level probe is shown in fig. 7.11a. An insulated rod is inserted into the tank. In practice, a metal rod coated with PVC or PTFE is used to prevent corrosion. The capacitance between the rod and the tank walls has two components: C_1 above the liquid (or solid) surface and C_2 below. These capacitances depend on the geometry of the installation (e.g. rod diameter and the distance to the wall) and the dielectric constants of the liquid and the vapour above the surface.

As the liquid level rises, C_1 will decrease and C_2 will increase in value. The two capacitors are effectively in parallel, and as liquids and solids have a higher dielectric constant than vapours, the net result is an increase in capacitance with level.

The change in capacitance is, however, small, so the probe is usually used in conjunction with a measuring bridge/amplifier circuit, as in fig. 7.11b. Normally a drive frequency of around 100 kHz is used. The detection circuit usually needs to be fairly close to the probe (typically 50 m) to prevent the signal change in capacitance being swamped by the capacitance of the connecting cable.

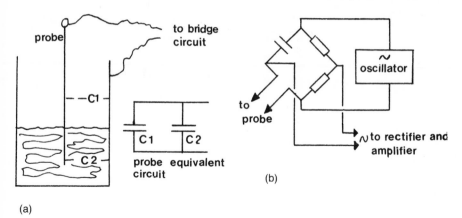

Fig. 7.11   Capacitive probe method. (a) Physical arrangement. (b) AC bridge.

The physical construction of the probe and tank is critical. Obvious errors will occur if the tank sides are not parallel to the probe, and the simple arrangement of fig. 7.11a will obviously not work with non-conducting tanks (e.g. glass fibre or plastic). In these applications the capacitors may be formed from two parallel rods, or constructed as a rod and concentric tube.

The capacitance is not only dependent on level, but is also affected by change in the value of the dielectric constant of the liquid and/or vapour. If an uncoated probe is used, changes in the resistance of the liquid will also be perceived as changes in level. The transducer will only work satisfactorily with liquids or solids with relatively high dielectric constant. Liquids that are frothing, foaming or boiling have a low dielectric constant (typically less than 2), which is also probably variable.

Capacitance probes work with a wide range of liquids and solids, and are probably the most versatile of all level sensors. They do, unfortunately, have a rather bad reputation from early units which were prone to drift and suffered from difficult installation requirements (e.g. measuring circuits within a metre of the probe). The more serious objection still remains, however, in that the probe has to be designed specifically for each and every application with regard to tank geometry, dielectric constant, and so forth. A capacitance level probe is not a universal transducer in the way that a thermocouple or pressure transducer is.

Capacitance probes are probably the best solution to the

measurement of bulk solid level (with the possible exception of direct weighing). Care must be taken, though, to avoid errors due to changes in dielectric constant from moisture or material compaction (the density of a bulk powder tends to increase towards the bottom of a solid). An unexpected hazard which can damage the measuring circuits is the build-up and discharge of high-voltage static electricity inside silos.

7.4.2. Resistive probes

If the resistance of the liquid is reasonably consistent, level can be inferred by observing the resistance between two metal rods inserted into the liquid. This requires uncoated probes, usually of stainless steel to overcome corrosion problems. Although a DC supply can, in theory, be used, in practice an AC supply is used to prevent electrolysis and plating effects. A bridge measuring circuit (similar to fig. 7.11b but with resistance rather than capacitance) is usually employed local to the probe. The circuit is, in reality, measuring a combined resistance/capacitance impedance for the probe. Resistive probes are less versatile than capacitance probes, and tend to be found mainly in level sensing (on/off) applications.

7.5. Ultrasonic methods

Ultrasonic methods are based on high-frequency sound waves produced by the application of a suitable AC signal to a piezo-electric crystal. A typical device is shown in fig. 7.12. Operation at frequencies up to 1 MHz is feasible, but 50 kHz is more typical in industrial applications (the high frequencies are used in, for example, medical ultrasonic scanners). The operating frequency is chosen to correspond with the resonant frequency of the transmitter and receiver.

The principle of ultrasonic level measurement is shown in fig. 7.13. An ultrasonic transmitter and receiver are placed at the top of the tank, and an ultrasonic beam is directed at the surface of the liquid (or solid) contents. The level in the tank is inferred from the reflected signals.

The reflected signal will be a delayed version of the transmitted signal, with a delay of $2d/v$ seconds, where d is the distance of the

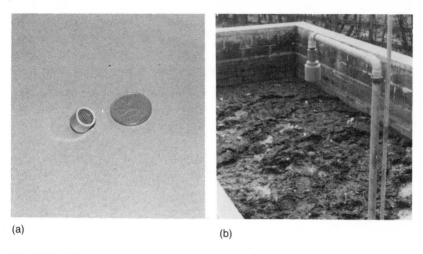

(a)

(b)

Fig. 7.12 (a) Ultrasonic transmitter. (b) Ultrasonic measurement of sludge level. (Photo courtesy of Hycontrol Ltd.)

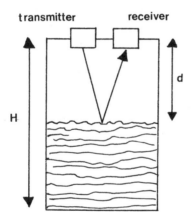

Fig. 7.13 Principle of ultrasonic level measurement.

surface from the tank top, and v is the velocity of sound in the medium above the surface (which need not be air; the principle could be applied to non-miscible liquids, or to determine the position of a liquid/sludge interface. Maritime variations are, of course, Asdic and depth sounders).

The velocity of sound in air, for example, is about 3000 m s^{-1}. For a level which can vary from 0 to 10 m, the delay will be 0 to 70 ms. There are two ways of measuring the delay time, the choice being determined by the range of measurement compared with the wavelength of the sound. If d is large (as it usually is in level measurement), a narrow sound pulse is transmitted as either a sound burst or a narrow spike. The receiver will see two pulses: one direct from the transmitter and one reflected off the surface. The time delay can be measured directly as shown in fig. 7.14a, and the position of the surface computed directly.

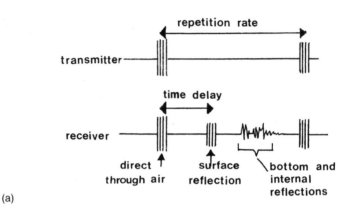

(a)

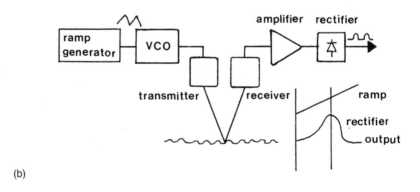

(b)

Fig. 7.14 Ultrasonic techniques. (a) Pulsed transmitter. (b) Swept oscillator method for small distances.

The accuracy of the pulsed method decreases as the delay approaches the pulse width. The arrangement of fig. 7.14b is then used. A broadband transmitter and receiver are used, with the transmitter frequency swept by a voltage-controlled oscillator. As before, the receiver receives two signals, one direct and one reflected. At one particular frequency where the difference in path length is an exact multiple of the wavelength, the two signals will combine to give a peak. By noting the wavelength at which this peak occurs, the reflected path distance can be computed from the formula:

$$2d = v/f \qquad (7.15)$$

where d is the distance to the surface, v the velocity of propagation, and f the frequency at which the peak occurs. Equation 7.15 gives distances of a few millimetres, so this technique is usually found in thickness gauges and medical instruments. Note also that the result is ambiguous, as peaks will be seen at multiples of the wavelength.

Both methods of measurement rely on a knowledge of the velocity of propagation. This varies according to the material (1440 m s^{-1} for water, 5000 m s^{-1} for steel) but also, for gases, with pressure and temperature. The velocity in air, for example, varies by about 1% for a 30°C change in temperature. If environmental changes are likely to cause errors, these external effects can be measured separately and suitable compensation applied. Alternatively, a second calibration receiver can be located a known distance from the transmitter to measure v directly.

A less obvious source of error can arise from unwanted multiple reflections from walls. If the position of the transducer is correctly situated, these will occur *after* the surface reflection. Multiple secondary reflections determine the maximum pulse rate for systems based on fig. 7.14a.

7.6. Nucleonic methods

7.6.1. Principles

Radioactive isotopes (such as cobalt 60) spontaneously emit gamma or beta radiation. As this radiation passes through material, it is attentuated according to the relationship:

$$I = I_0 \exp(- \mu \varrho d) \qquad (7.16)$$

where I_0 is the initial intensity; μ is a constant for the material, called the mass attenuation coefficient; ϱ is the density of the

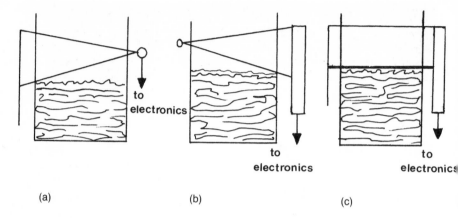

(a) (b) (c)

Fig. 7.15 Radioactive source level measurement. (a) Line source, point detector. (b) Point source, line detector. (c) Collimated line source, line detector.

material; and d is the thickness of material between the source and the measuring point.

Equation 7.16 allows a conceptually simple level transducer to be constructed, as in fig. 7.15a. A strip radioactive source is placed on one side of the tank, and a point detector is placed on the other side. Gamma radiation above the liquid (or solid) surface will be slightly attentuated by the air and tank walls. Radiation passing through the liquid will be greatly attenuated. The intensity of the radiation seen by the detector will therefore be related to the liquid level in the tank (being a maximum when the tank is empty).

Figures 7.15b and c show variations on the same idea. Figure 7.15b is a point source and linear detector and fig. 7.15c is a linear source and detector. Figure 7.16 is a typical installation, used to measure liquid steel level on a steel casting plant.

The strength of a radioactive source decays exponentially with time, and is described by its 'half-life', which is the time for the source strength to fall to half its initial value. Nucleonic level sensors therefore have a 'built-in' span drift, and consequently require periodic adjustment. A common source is cobalt 60. This has a half-life of 5.3 years and will exhibit a 1% change in about a month. Sources are normally usable (with readjustment of the sensitivity) for about one half-life. Other common isotopes are caesium 137 (half-life 37 years) and americium 241 (half-life 458 years).

Fig. 7.16 Radioactive source level measurement in steel casting. The source is in the large housing to the left of the mould, and the GM tube sensor on the right.

Nucleonic level sensors have many advantages, and in some applications they are the only solution to a level measurement problem. Because no contact with the container is needed, nucleonic sensors can be used with liquids or solids regardless of temperature, pressure, or corrosion effects.

There are some disadvantages, of course. Nucleonic systems have to be designed individually, and the sources and detectors chosen to suit the size of tank and the attenuation coefficient of the material being measured. Safety and legislation aspects are discussed in section 7.6.3.

7.6.2. Radiation detectors

Level measurement systems using radioactive sources require some form of detector to convert the attenuated radiation to an electrical signal. Two detectors are commonly used, the Geiger–Müller (GM) tube and the scintillation counter.

The GM tube is constructed with a sealed cylinder containing an axial wire anode, as shown in fig. 7.17a. The cylinder is earthed and the wire kept at a high positive potential (typically 300–500 V). The tube is filled with a halogen gas. As a radioactive particle passes through the tube, the gas is ionised forming electrons and positive ions which are attracted, respectively, to the wire and the tube wall. Collisions between the electrons and the gas release secondary electrons, and an avalanche effect occurs where a discharge occurs along the whole length of the central wire. This produces a negative voltage pulse at the anode. The pulse is of constant amplitude. Because a pulse occurs for each particle traversing the tube with sufficient energy to start the avalanche, the pulse rate is dependent on the strength of the radiation.

The pulse rate is also dependent on the voltage of the anode, as shown in fig. 7.17b. There is, however, a region called the 'plateau', where the tube is relatively insensitive to changes in anode volts. GM tubes are normally operated at a voltage in the centre of the plateau.

Scintillation counters detect ionising particles by their passage through a crystal. When an ionising particle passes through an material, atoms are raised to an excited state, subsequently emitting a short pulse of light as they return to their ground state. Most materials reabsorb this light pulse again. Scintillation crystals, however, are transparent to their own excited radiation, so the passage of ionising radiation can be observed as short pulses of light. Common crystals are NaJ(T1) and ZnS(Ag) phosphors.

The light pulses are very weak, so the construction illustrated in fig. 7.17c is used to detect them. The scintillation crystal is 'viewed' by a photomultiplier tube (see section 8.4.5) which gives random pulse outputs whose average frequency depends on the received radiation. Like the GM tube, the photomultiplier needs a stabilised HT supply.

Both the GM tube and the scintillation counter give a semi-random pulse train whose average repetition rate is radiation dependent. This must be converted to a DC voltage before it can be used to indicate radiation strength and hence liquid, or solid, level. One common way of achieving this is shown on fig. 7.17d. The narrow pulses from the primary sensor are used to fire a monostable to give broader, fixed width, pulses. The output from the monostable is smoothed by an Op Amp filter circuit to give a DC signal which is related to the radiation intensity.

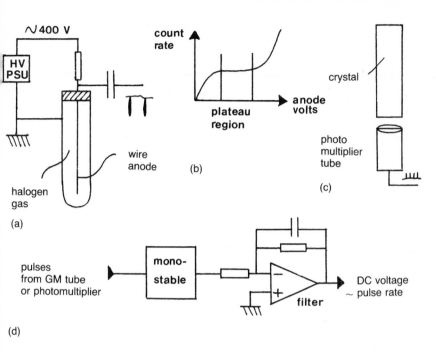

Fig. 7.17 Radiation detectors. (a) Geiger–Müller (GM) tube. (b) Operating voltage for GM tube. (c) Scintillation counter. (d) Detection circuit.

7.6.3. Safety and legislation

The most serious objection to the use of radiation sources is undoubtedly the inherent health hazard, which brings as a side-effect a plethora of legislation covering their use, storage, transportation and disposal. The user must accept that radioactive sources are dangerous, and treat them with respect. In the UK, sites using sources have to be registered, and records kept for many statutory bodies including the Factory Inspectorate, Health and Safety Executive, Department of the Environment, Radiochemical Inspectorate, Department of Health and Social Security and the National Radiological Protection Board. To this army of legislative might is added the inevitable suspicions of personnel in the area of the application. The use of radioactive sources brings more clerical and public-relations problems than technical difficulties.

Almost all sources used in industry are sealed, i.e. the source is contained permanently in an enclosure which can be totally sealed when not in use (unsealed sources are used in medical and laboratory applications and are subject to different legislation). When not in use, a shutter is opened allowing a collimated beam to be emitted (typically $40° \times 7°$).

Source strength is determined by the number of disintegrations per second (d.p.s.) at the source. Two standards are in common use; the curie, Ci (3.7×10^{10} d.p.s.) and the bequerel, Bq (1 d.p.s.). It follows that 1 mCi = 37 MBq. A typical industrial source would be 500 mCi (18.5 GBq) caesium 137, although sources up to 5 Ci (185 GBq) are used occasionally.

The biological effects of radiation are complex, and there is no real 'safe dose'. The legislation is, in general, based on the concept of 'as low as is reasonably achievable' (ALARA) and 'acceptable levels' of biological effects. It is significant that the genetic and carcinogenic effects of radiation exhibit no lower threshold, and all exposure can be considered potentially harmful. Legislation has become noticeably more restrictive over the years.

The absorbed dose (AD) is the energy density absorbed. Two units are, again, used; the millirad (Mrad) and the milligray (mGy):

$$1 \text{ mrad} = 6.25 \times 10^4 \text{ MeV g}^{-1}$$
$$1 \text{ mGy} = 6.25 \times 10^6 \text{ MeV g}^{-1}$$

It follows that 1 mGy = 100 mrad.

The absorbed dose is not, however, directly related to biological damage as it ignores the differing effects of α, β and γ radiation. A quality factor, Q, is defined (20 for α, 1 for β, γ) which allows a dose equivalent, DE, to be calculated from:

$$DE = Q \times AD \tag{7.17}$$

There are, yet again, two units in common use: millirem (mrem) and millisievert (mSv). These are defined by:

$$\text{mrem} = \text{mrad} \times Q \tag{7.18}$$

$$\text{mSv} = \text{mGy} \times Q \tag{7.19}$$

It again follows that 1 mSv = 100 mrem.

The AD and DE represent energy density absorbed, so the strength of radiation, the dose rate, is time related and expressed in mrad h^{-1}, mGy h^{-1}, mrem h^{-1} or μSv h^{-1}. Although it is possible to calculate dose rate from source strength, it is usually more convenient to refer to manufacturers' data sheets. A cobalt

60 source, for example, gives 3.39 mGy h^{-1} per MBq at 1 cm. Dose rates fall off as the inverse square of distance: doubling the distance gives one-quarter the dose rate.

Medical effects are related to both the dose rate and the total lifetime dose (DE). It follows that there are two aspects to protection from radiactive sources: the instantaneous dose rate and the dose equivalent (the cumulative effect) must both be controlled.

For industrial users, the legislation recognises three classes of person:

(i) Classified workers who are trained personnel wearing dose monitoring devices (usually film badges). These workers are allowed dose rates up to 2.5 mrem h^{-1} and an annual dose up to 5 rem. Medical records and dose history must be kept.

(ii) Supervised workers who operate under the supervision of classified workers, and are permitted dose rates of 0.75 mrem h^{-1} and an annual dose of 1.5 rem.

(iii) Unclassified personnel: this refers to other workers and the general public, for whom permitted dose rates are 0.25 mrem h^{-1} and the annual dose is 0.5 rem.

The legislation emphasises that the above are *not* design criteria, and the ALARA principle should apply. Dose rates are dependent on source strength, shielding and distance, and total dose on dose rate and time of exposure.

The legislation covering radioactive sources is lengthy and complex, so this discussion should only be regarded as a very brief summary. Although the underlying theory of nucleonic level and thickness transducers is straightforward, the legal aspects are a minefield for the unwary. Professional advice should always be sought at the design stage.

7.7. Level switches

Many level measurement and control applications involve 'surge' tanks where the level varies to cope with a sudden rise, or fall, of supply or demand. In these situations, tight control is not needed and would even negate the purpose of the tank by directly coupling inflow and outflow. All that is usually needed is an indication that the level has reached, or fallen to, some predetermined level.

Any of the previous transducers can be used as a level switch, but simpler devices are available. The commonest are simple float-operated switches which will work with most liquid/tank combinations. Solids can be detected by horizontally mounted capacitance probes. Other alternatives are rotating paddles or vibrating reeds which 'seize' when immersed and operate a microswitch.

The above sensors have moving parts which are prone to jamming when contaminated. Two non-moving alternatives are shown on fig. 7.18. The heated probe contains a heater and a temperature sensor. The heat loss will be greater (and hence the temperature lower) when the probe is immersed. The light-reflective probe uses total internal reflection (see section 8.3.2) to detect immersion of the plate.

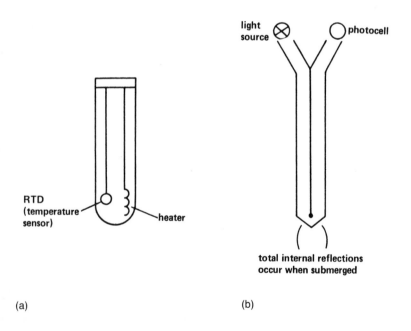

(a)   (b)

Fig. 7.18   Unusual level switches. (a) Level switch working on heat loss from submerged object. (b) Optical level switch.

Chapter 8
Optoelectronics

8.1. Introduction

Optical devices are becoming increasingly common in instrumentation and control. Such devices are covered by the multidiscipline term of optoelectronics. Optical sensors range from simple measurement of light intensity (photometry) to non-contact transducers for the measurement of position or temperature. Light emitters (such as LED or LCD devices) are used to display data to operators. Data transmission via fibre optic cable gives a high-speed alternative to electrical and pneumatic signals with almost perfect noise immunity and no explosion risk in hazardous atmospheres.

Optoelectronics is therefore a very broad subject, spanning across lasers, lenses, instrumentation, electronic circuits and the physics of electromagnetic radiation. This chapter covers the principles behind the optoelectronic devices that may be encountered by a process control engineer.

8.2. Electromagnetic radiation

The term optoelectronics implies the use of visible light. In reality, light is energy in the form of electromagnetic (EM) radiation, identical in nature to radio waves, gamma rays, infrared radiation and X-rays. A full description of EM radiation would require a fairly detailed knowledge of theoretical physics, and as such is beyond the scope of this book. Users of optoelectronics, however, do not need a deep understanding of the underlying theory.

Electromagnetic radiation propagates through space in a

manner somewhat analogous to waves propagating through a liquid. Any EM radiation can therefore be described by its *frequency* (i.e. the number of oscillations per second passing a fixed point), its *wavelength* (the distance between successive maxima) and the *velocity of propagation*. These are related by:

$$c = \lambda f \qquad (8.1)$$

where c is the velocity in metres per second, λ is the wavelength in metres and f is the frequency in hertz.

The velocity of EM radiation is constant for all frequencies, but is dependent on the transmission medium. The velocity is highest in a vacuum, for which c is 2.998×10^8 m s^{-1}. The velocity in glass is about 1.9×10^8 m s^{-1}.

The ratio of the EM velocity in vacuum to the velocity in a medium is defined as the *index of refraction* of the medium, i.e.

$$\mu = \frac{c}{v} \qquad (8.2)$$

where μ is the index of refraction, c is the velocity of EM radiation in vacuum, and v is the velocity of EM radiation in the medium. The index of refraction for glass is about 1.57.

Electromagnetic radiation is described by its frequency or wavelength. The range is large, as shown by the EM spectrum of fig. 8.1, which also demonstrates that the only difference between radio waves and visible light is the frequency of the radiation. Although EM radiation can be described by frequency or wavelength, it is usual to describe radiation used in microwave or broadcasting by its frequency, and radiation in optoelectronics by

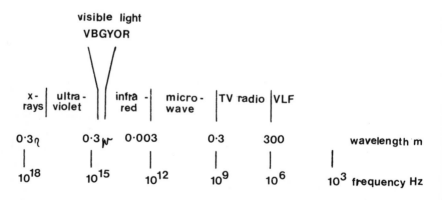

Fig. 8.1 The electromagnetic spectrum.

its wavelength. VHF radio, for example, is described as having a frequency between 30 and 300 MHz, whereas visible light is described as the small range with a wavelength between 0.38 μm and 0.78 μm.

Optoelectronics is concerned with EM radiation with wavelengths in the range 0.3 μm to 30 μm, i.e. from just inside the ultraviolet range, through the visible light region to well into the infrared. In the rest of this chapter we will use the loose term 'light' for convenience, although the actual range of EM radiation is larger.

The wavelength of EM radiation is often expressed in Ångstrom units (Å) which is defined as 10^{-10} m. Red light therefore has a wavelength of 7800 Å and violet a wavelength of 4000 Å.

Visible light is perceived as having colour. Figure 8.1 shows that the colour is determined by the wavelength. Red light has the longest wavelength at 0.78 μm, with violet at 0.4 μm. The range from 0.8 μm to 100 μm is called the infrared, and that from 0.01 μm to 0.4 μm the ultraviolet.

8.3. Optics

8.3.1. Reflection and mirrors

Most optoelectronics devices incorporate mirrors and lenses to gather and focus light. In this section a simple description of the devices is given.

For most practical purposes, light can be considered to travel in straight lines. This can be demonstrated by the experiment in fig.

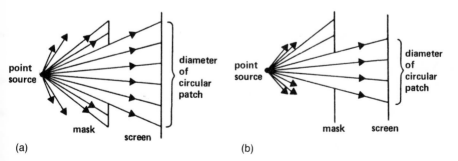

(a)

(b)

Fig. 8.2 Demonstration of the straight-line nature of light. (a) Large hole. (b) Small hole.

8.2, where the hole in an opaque screen is progressively reduced, resulting in a corresponding reduction in size of the circle of light on the screen. (If the hole size is reduced below 0.01 mm, other effects start to occur and a diffuse patch of light is seen. Provided we are not concerned with the passage of light through very small apertures, these effects need not concern us.) If light travels in straight lines, the analysis of optical devices is simplified considerably by considering light to be a collection of light rays and looking at what happens to a few of them.

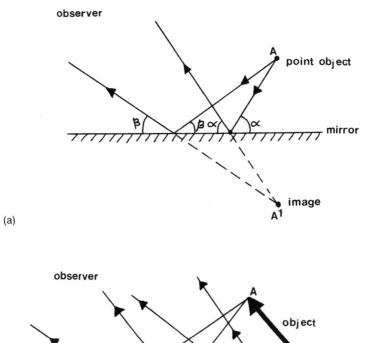

(a)

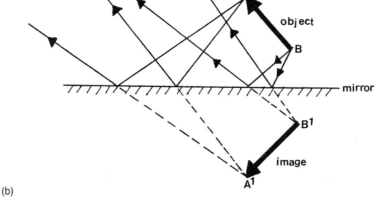

(b)

Fig. 8.3 Reflections from a mirror. (a) Point object. (b) Finite object.

In fig. 8.3a, for example, *A* is a point object in front of a mirror. The object will emit light in all directions, so there will be a myriad of light rays. To analyse the effect of the mirror we need only consider any two, as shown in fig. 8.3a. These strike the mirror at angles α and β, and are reflected with the same angles as shown. To an observer, the reflected rays appear to come from a point *A*, the same distance behind the surface as *A* is in front.

Figure 8.3b illustrates the general case of an object of non-point size. By considering two rays from each end of the object, it can be seen that an image A^1, B^1 is formed apparently behind the mirror.

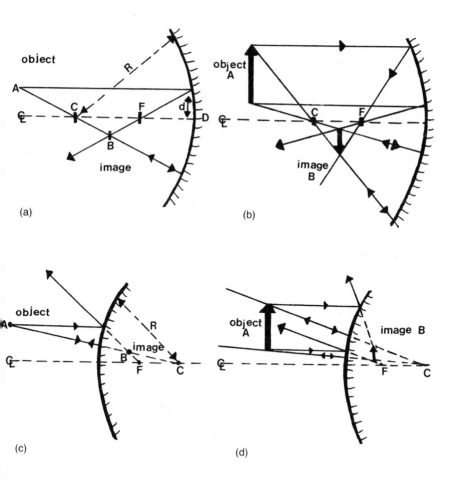

Fig. 8.4 Reflections from curved mirrors. (a) Point object, concave mirror. (b) Finite object, concave mirror. (c) Point object, convex mirror. (d) Finite object, convex mirror.

Figure 8.4a shows a curved concave mirror with centre of curvature at point C and radius R. To examine its action we consider two rays of light, one parallel to the centre line of the mirror and one to the centre of curvature of the mirror. The ray parallel to the centre line is reflected through point F where F_D is $R/2$ providing D is small compared with R. The ray through the centre of curvature of the mirror will be reflected back along its own path as shown. A small object A will therefore form an image at point B.

The distance from the mirror to the focus, F_D, is called the focal length. Any ray parallel to the centre line of the mirror will pass through point F, which is correspondingly known as the focus of the mirror. Obviously, any light originating at point F will emerge as a parallel beam, a property used, for example, in searchlights and car headlamps.

The behaviour of light from a finite-sized object can be analysed by considering two rays from each end as shown in fig. 8.4b. An inverted image of the object is produced at B. If a small screen is placed at this point an image will be seen. It also follows that an object placed at B will produce an image at A.

Figure 8.4c shows the analogous case for a convex mirror with, again, centre of curvature C and radius R. As before, by considering the action of a ray parallel to the centre line and one to the centre of curvature, it is found that an image is formed at point B *behind* the mirror. Point F is, again, the focus of the mirror. Figure 8.4d illustrates the formation of an image from a finite-sized object. The image is, again, formed behind the screen.

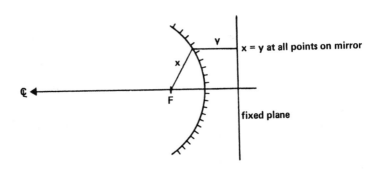

Fig. 8.5   Construction of a parabolic mirror.

There is a fundamental difference in the images of figs. 8.4b and 8.4d. In fig. 8.4b the image is formed by convergence of light rays at a point. Such an image can be focused on to a screen and is called a *real* image. In fig. 8.4d, the rays *appear* to diverge from a point and the image cannot be focused on a screen. Such an image is said to be a *virtual* image.

The analysis of fig. 8.4 is based on mirrors with a circular section and a diameter that is small by comparison with the radius. If these conditions are not met, the image will be distorted. A parabolic mirror, however, will behave as described regardless of the diameter. The focus of the mirror is at the focus of the parabola, as shown in fig. 8.5. A parabolic mirror is, of course, more expensive to manufacture.

8.3.2. Refraction and lenses

The speed of light is dependent on the material through which it travels as described by the index of refraction defined by equation 8.2. One consequence of this speed change is that the direction of a ray of light changes as it passes from one medium to another (e.g. from air to glass). If the light passes from an optically less dense medium (e.g. air to glass), the ray bends towards the normal. If the light passes from a more dense medium (e.g. glass to air), the ray bends away from the normal. Both these cases are illustrated in fig. 8.6a.

The behaviour of the ray is described quantitatively by Snell's law which states that:

$$\frac{\sin \alpha}{\sin \beta} = \frac{\text{Velocity of light in medium 1}}{\text{Velocity of light in medium 2}} = \frac{\mu_2}{\mu_1} \qquad (8.3)$$

If medium 1 is air, the value of μ_1 is unity for all practical purposes so:

$$\frac{\sin \alpha}{\sin \beta} = \mu \qquad (8.4)$$

where μ is the refractive index of the denser medium. Angle α is called the angle of incidence, and angle β the angle of refraction.

Figure 8.6b shows light passing from glass to air at increasing angles. Eventually at an angle θ the refracted ray just skirts the surface of the glass. At angles greater than θ, a ray of light will be internally reflected back into the glass. The angle θ is called the critical angle. The value of θ is given by:

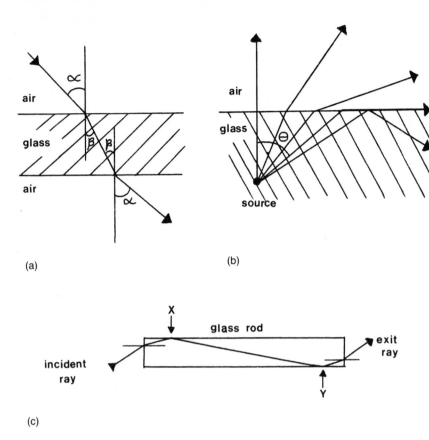

(a)

(b)

(c)

Fig. 8.6 The refraction of light. (a) Principles of refraction. (b) Total internal reflection. (c) Conveying a light beam down a glass rod. Total internal reflection occurs at points X and Y.

$$\sin \theta = 1/\mu \qquad (8.5)$$

For glass, with a μ of 1.57, the value of θ for a glass/air interface is about 40°.

Internal refraction allows a light beam to be conveyed along a glass tube as shown in fig. 8.6c, which is the basis of fibre optic light guides (described further in section 8.6).

Lenses are based on the phenomenon of refraction. Figure 8.7a shows a convex lens. A light ray entering on one side will exit in a different direction. The lens surfaces are shaped such that any ray parallel to the axis of the lens passes through the same point called, not surprisingly, the focus of the lens. The distance from the lens to the focus is called the focal length of the lens.

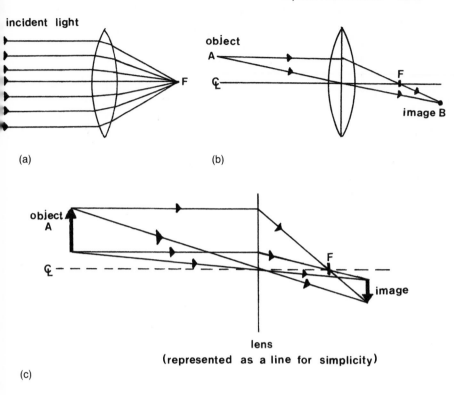

Fig. 8.7 The convex lens. (a) The action of a lens. (b) Formation of an image of a point object. (c) Formation of an image of a finite object.

The action of a lens can be analysed by considering two rays from a point in a similar way to that demonstrated for mirrors (fig. 8.7b). A ray parallel to the axis of the lens will pass through the focus, and a ray through the lens centre will not be diverted (because entry and exit angles are equal). An object at point A will produce an image at point B. The image of a finite-sized object can be found by considering rays from each end (fig. 8.7c). Under the conditions shown a real image is formed.

The position and size of the image can be found by simple geometry. Referring to fig. 8.8, it follows that:

$$d = DF/(D - F) \tag{8.6}$$

and

$$h = dH/D \tag{8.7}$$

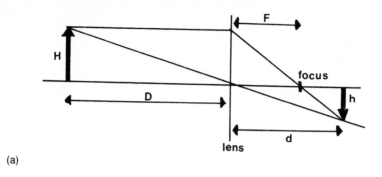

(a)

(b)

Fig. 8.8 Analysis of the operation of a convex lens. (a) Mathematical analysis of image formation. (b) Magnifying operation.

Note that if $D = 2F$, $h = H$. If D lies between $2F$ and F, h is larger than H (a magnified image is produced). If D is less than F, the rays after the lens diverge as shown in fig. 8.8b. Under these circumstances a virtual image is formed on the same side of the lens as the object (this is the basis of a hand-held magnifying glass).

Light rays diverge from a concave lens, but the action can be analysed in a similar manner. Figure 8.9a analyses two rays from a point source and fig. 8.9b a finite object. A concave lens always produces a reduced virtual image.

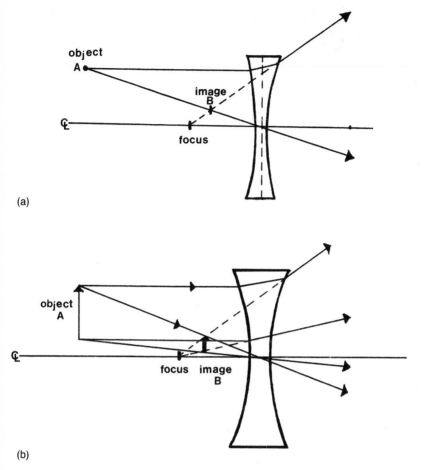

Fig. 8.9 Operations of a concave lens. (a) Formation of image from a point object. (b) Formation of image from a finite object.

8.3.3. *Prisms and chromatic aberration*

The index of refraction, μ, of glass was given in section 8.2 as 1.57. In practice μ is found to vary slightly with the wavelength of light, being greater towards the red end of the spectrum. This means that violet light is bent more on crossing an air/glass interface.

White light is a combination of all the colours of the spectrum, so a ray of white light will be split into its constituent colours on passing from air to glass (fig. 8.10a). The effect can be increased by a prism (fig. 8.10b). The splitting of light into its constituent colours is called spectroscopy, and can be used for chemical analysis, a topic covered further in section 8.9.1.

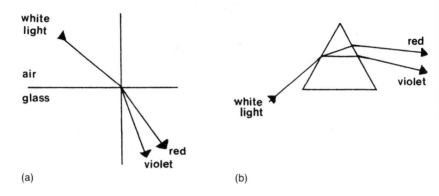

Fig. 8.10   The variation of refractive index with wavelength. (a) The effect of passing white light from air into glass. (b) A prism.

The variation of the refractive index with wavelength causes the focal length of a lens to be colour dependent, as shown in fig. 8.11a. This is known as chromatic aberration and appears in

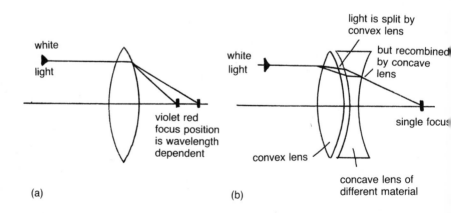

Fig. 8.11   Chromatic aberration. (a) The effect of chromatic aberration. (b) The cure.

cheap optical instruments as a coloured haze around objects.

Chromatic aberration can be almost eliminated by combining a convex and concave lens (fig. 8.11b). The two lenses are made of glass with different refractive index. By careful design, the chromatic aberration of the two lenses can be made to cancel, giving a focal point independent of wavelength.

8.4. Sensors

8.4.1. Photoresistors

An optoelectronic sensor is a device which converts a light signal into an electrical signal. The cheapest, and easiest to apply, sensor uses light to change the resistance of a semiconductor and is accordingly called a photoresistor.

The resistance of a semiconductor is determined largely by the number of electrons in the conduction energy band. Normally a few electrons are excited across the energy gap from the valence band to the conduction band, but the resistance of a semiconductor at room temperature is quite high.

If light of sufficient energy is absorbed by the semiconductor, more electrons are excited and can pass from the valence band to the conduction band, creating hole (in the valence band) and electron (in the conduction band) pairs. The absorbed light creates pairs in proportion to the intensity, causing a decrease in

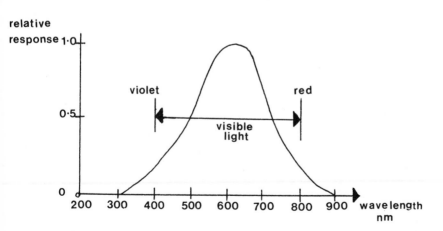

Fig. 8.12 Response of cadmium sulphide photoresistor.

the semiconductor resistance.

The wavelength of the light is rather critical. If the wavelength is too long, the energy is too low to create the electron/hole pairs. If the wavelength is too small, it will all be absorbed at the surface and cause no net change in resistance. The spectral response of a typical photoresistor is shown in fig. 8.12.

Useful photoresistive materials are:

Material	Energy gap (eV)	Range (Å)	Peak sensitivity (Å)
Cadmium sulphide	2.54	4 000– 8 000	5 200
Cadmium selenide	1.74	6 800– 7 500	7 000
Lead sulphide	0.4	5 000–30 000	20 000
Lead selenide	0.3	7 000–58 000	40 000
Indium antimonide	0.16	6 000–70 000	55 000

Visible light, it will be remembered, covers the range 4000 Å to 8000 Å. Photoresistors cover the range from visible light to well into the infrared. The commonest of the above materials is cadmium sulphide (CdS).

A typical device is the CdS-based ORP12 shown in fig. 8.13. The semiconductor is arranged in a thin layer to give a large surface area and minimum thickness. The change in resistance is large; an ORP12 will go from around 2M in the dark to around 100R in normal room lighting. This large change allows simple

Fig. 8.13   ORP12 photoresistor.

circuits to be used.

The resistance change is, however, very non-linear, and the actual resistance value is dependent on temperature as well as light intensity. It is therefore important to minimise dissipation in the device. Photoresistors also have a significantly large time constant: typical values range from about 0.1 ms for a lead sulphide cell to over 100 ms for a cadmium sulphide cell.

8.4.2. Photodiodes

A photodiode consists of a back-biased *p–n* junction which, under dark conditions, behaves as a normal back-biased diode. With these conditions the only current flowing through the diode of fig. 8.14a will be the usual leakage current (typically 1 μA).

Absorbed light will generate electron/hole pairs as described above, and the current through the diode will increase to a typical value of 100 μA. A photodiode has the response similar to fig. 8.14b which shows that it can be considered as a constant current device with the current determined by the light intensity.

The current/intensity relationship is quite linear, and the response is fast; typically 0.2 μs but devices as fast as 1 ns are available. In general, photodiodes are the smallest optical sensor which, in conjunction with their high speed, makes them well

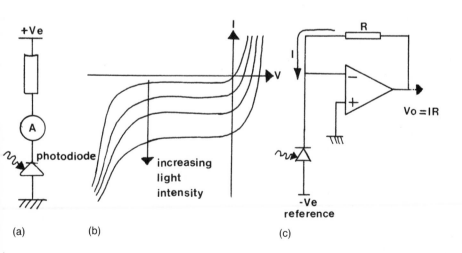

Fig. 8.14 The photodiode. (a) Basic circuit. (b) V/I/light relationship. (c) Practical circuit.

suited for fibre optic data transmission and similar applications. Typical operating wavelengths are 8000 Å to 11 000 Å (silicon) and 13 000 Å to 20 000 Å (germanium).

The relatively low level current can easily be converted to a high-level voltage using a DC amplifier, as in fig. 8.14c. The light-dependent diode current flows through R to give an output voltage IR which is directly related to light intensity.

8.4.3.   Phototransistors

A phototransistor is a variant of the photodiode, and is shown schematically in fig. 8.15a. A photodiode is connected between collector and base so that the light-dependent diode current becomes the transistor base current. The transistor multiplies this by its current gain, β, to give a far larger collector/emitter current change.

Unfortunately the dark current is also multiplied by β. A typical device has the response of fig. 8.15b, with a change from dark to light of 0.1 mA to 10 mA.

Using similar ideas to fig. 8.15a, it is also possible to construct photothyristors, photoFETs, and photoDarlingtons, but these are relatively rare.

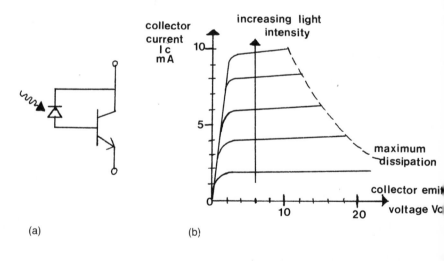

(a)   (b)

Fig. 8.15   The phototransistor. (a) Construction. (b) V/I/Light relationship.

8.4.4. Photovoltaic cells

A photovoltaic cell is a *p–n* junction which is designed to generate a voltage when light is absorbed. The mechanism by which this occurs is complex, and a full description is beyond the scope of this book. The voltage generated is small; typically 0.4 V in sunlight for a silicon cell, with an available current between 30 mA and 100 mA. The power can be increased by series/parallel combinations, often called solar cells.

A photovoltaic cell can be used in two modes, open circuit voltage and short circuit current. In the open circuit voltage mode the voltage is logarithmically related to light intensity, which is useful in photometry applications where a large light range is to be covered. The short circuit current is linearly related to intensity and is used where increased accuracy is required over a small light range.

Photovoltaic cells are expensive and slow (with time constants similar to photoresistors). Their main application is in battery-less photometry and solar-powered circuits (a small solar cell can easily supply CMOS circuits).

8.4.5. Photomultipliers

The most sensitive optical sensor is the photomultiplier tube shown schematically in fig. 8.16a. At the input end a cathode coated with photoemissive material is held at a large negative voltage (typically 1 kV). At the output end of the tube, the anode is held near 0 V via the load resistor R.

Between the cathode and anode there are several intermediate electrodes called *dynodes*. There are four dynodes on fig. 8.16a for simplicity, but in practice many more are used. The dynode voltages are equally spaced between the anode and cathode, usually by a resistor chain as in fig. 8.16b.

When light strikes the photoemissive cathode, electrons are released in quantities dependent on the light intensity. They are attracted to the more positive dynode, D_1. The dynodes are coated with a material from which electrons are easily detached. When the electrons from the cathode strike D_1, more electrons are released, which are now attracted to D_2. The process is repeated along the tube, so the number of electrons arriving at the anode is a large multiple (typically 10^6) of those released by

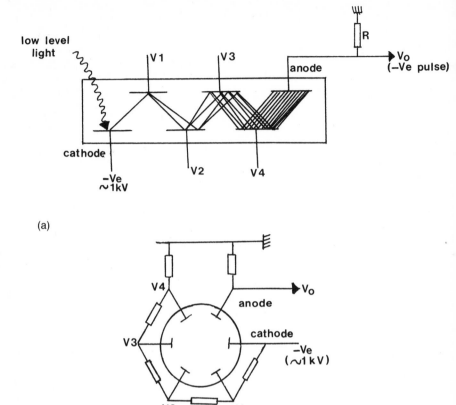

Fig. 8.16 The photomultiplier. (a) Construction and principle of operation. (b) Derivation of dynode voltages.

the light absorbed at the cathode.

The electrons arriving at the anode cause a current to flow through the load resistor R, giving a negative output voltage proportional to the (low-level) input light intensity.

Photomultipliers are used where very low light levels are to be measured. A typical application is in spectroscopy, a topic covered further in section 8.9.1.

8.4.6. Integrated circuit devices

All the basic light sensors require additional components to give a useful signal. It is increasingly common for semiconductor manufacturers to construct a complete photocell circuit (sensor and amplifier) in a single IC. Usually the IC incorporates a lens to focus the light on to the photosensor part of the circuit. Most of these, however, are designed to give an on/off digital indication (light above some threshold value/light below threshold value) rather than a photometry device with output proportional to light level.

8.5. Light emitters

8.5.1. Light-emitting diodes

Semiconductor light emitters are $p–n$ junctions that emit light when forward biased. As a $p–n$ junction forms a diode, the device conducts in one direction (emitting light) and blocks current in the reverse direction. Semiconductor light emitters are therefore usually called light-emitting diodes (LEDs).

The mechanism by which the light is produced is the recombination of electrons and holes between the valence and conduction bands, with the released energy appearing in the form of light. Fortunately it is not necessary to appreciate the underlying physics to use LEDs. The light emitted by an LED lies within a narrow band of wavelengths dependent on the materials used. Common devices are:

Material	Wavelength range (nm)
Gallium arsenide	890–980 (near infrared)
Gallium phosphide	530–580 (green to yellow)
Gallium phosphide/zinc oxide doping	620–700 (red)
Gallium arsenide/gallium phosphide	600–650 (orange)

The efficiencies of diodes based on these materials vary, but the actual perceived light intensity is more dependent on the eye's sensitivity to different colours. The maximum sensitivity is in the green/yellow/orange part of the spectrum, as can be seen in fig.

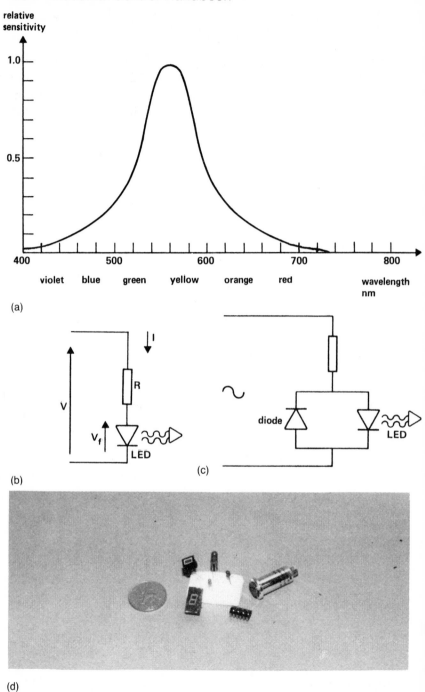

(a)

(b)

(c)

(d)

Fig. 8.17 Light emitting diodes (LEDs). (a) Spectral response of the human eye. (b) Basic circuit. (c) AC circuit. (d) Various LEDs.

8.17a. This is fortunate as gallium phosphide LEDs give a lower light intensity than other materials.

An LED is a current operated device, and operates with a relatively constant voltage of around 1.5–2.0 V. The operating current for most LEDs is around 20 mA. When operated from a voltage source, a current limiting resistor is needed (fig. 8.17b) with value:

$$R = \frac{V - V_f}{I} \tag{8.8}$$

The light intensity is approximately linearly related to the current, but this is not as apparent as might be thought. The eye's perception of intensity is logarithmic, so once an LED has attained a reasonable intensity, the apparent brilliance appears fairly independent of current.

Although an LED behaves as a diode, its reverse voltage is low (typically 5 V). If an LED is to be driven from an AC source, a protection diode should be used (fig. 8.17c).

LEDs are available in a variety of sizes and configurations, some of which are shown in fig. 8.17d. Although not as bright as incandescent bulbs, they operate at far lower currents. Seven-segment LED displays are widely used for numerical indication, a topic discussed further in chapter 7, Volume 2.

8.5.2. Liquid crystal displays

Liquid crystal displays (LCDs) are based on materials which can exhibit crystal-like structures in a liquid state. Such materials are normally transparent, but if an electric field is applied, complex interactions between the internal molecules and free ions cause the molecules to align and the liquid to become opaque. The advantage of LCDs is that they respond to an electric field and need no current. Operating power levels are correspondingly very low (typically about 0.1 mW per cm^2), which makes them very attractive for battery-powered devices.

The construction of an LCD cell is very simple, as shown in fig. 8.18a, consisting of two metallised glass plates separated by spaces. The gap between the plates is filled with the liquid. When an electric potential is applied to the plate, the liquid turns opaque.

An LCD requires an external light source. This can be

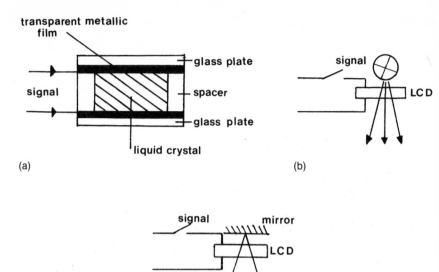

(a)

(b)

(c)

Fig. 8.18   The liquid crystal display. (a) Construction. (b) Transmissive mode.
(c) Reflective mode.

provided by an integral light emitter (called the transmissive
mode, fig. 8.18b) or by ambient light via a mirror (called the
reflective mode, fig. 8.18c).

An LCD will operate on DC, but in practice polarisation and
electrolysis effects give a greatly reduced life. It is usual for LCDs
to be run on pseudo AC, which gives greatly improved life
expectancy. A common way of providing AC drive is shown on
fig. 8.19a. A simple square wave oscillator is used to drive the cell
back plane and segment via an exclusive OR gate. Operating
waveforms are shown in fig. 8.19b. Using this arrangement a DC
component of less than 100 mV is achievable.

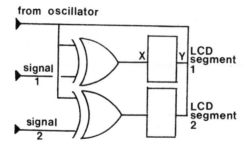

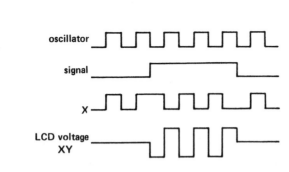

Fig. 8.19 Driving liquid crystal displays. (a) Circuit diagram. (b) Waveforms.

8.5.3. *Incandescent emitters*

The commonest light source is the common light bulb used in domestic lighting, car headlamps, torches, etc. When a material is heated, EM radiation is emitted (see section 2.6), and if the temperature is sufficiently high, part of the radiation will be in the visible part of the spectrum. The radiation, however, covers a large wavelength band far into the infrared, so only a small part of the input energy appears as visible light.

Incandescent emitters (or bulbs!) are inefficient, generate heat and have a limited life (which can be extended considerably by under-running, or by passing lamp-warming current through the filament in the off state to prevent thermal shock). They also have the undesirable characteristic of taking a large inrush current when first turned on. Their advantages are their simplicity and much higher intensity than any other emitter.

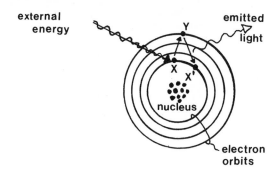

Fig. 8.20 Atomic sources of light.

8.5.4. Atomic sources

Figure 8.20 shows an atom which can be considered as consisting of a nucleus surrounded by shells of orbiting electrons. The electron shells are at fixed distances from the nucleus. If external energy is supplied, an electron can be excited from an orbit to a higher energy orbit (from, say, X to Y). The electron will only stay in this state for a short time before resuming its lower energy orbit. In falling back it loses energy which is emitted as EM radiation. In atomic sources of EM radiation, therefore, energy is input to change the orbits of electrons and the energy re-emerges as EM radiation. The wavelength of the radiation depends on the energy difference between electron orbits, so for any given material there are several possible transitions, each with its own characteristic wavelength (a topic discussed further in section 8.9.1).

In a neon lamp, for example, the energy is provided by the electric current flowing through ionised gas. For neon, the major transition produces light in the orange part of the spectrum giving the characteristic reddish/orange neon light.

In domestic fluorescent tubes the major transition produces ultraviolet radiation which is absorbed by the tube coating. Electrons in the coating are excited by ultraviolet radiation, but visible light is emitted when they resume their low-energy state.

Phosphorescence occurs in some materials and is used for luminous paint. The absorption of EM radiation in the form of light can excite electrons in some materials. Normally the time taken for an electron to resume its low energy state is very short (less than 10^{-8} s). In some materials, however, electrons can stay in a high-energy state for several minutes, slowly emitting light as they return to their normal orbits.

Many industrial displays are based on atomic light sources. Gas discharge displays, Nixies (a registered trade mark of the Burroughs Corporation) and phosphorescent displays are all variants of the above principles.

8.6. Fibre optics

Internal reflection of light was introduced in section 8.3.2. In fig. 8.21 a light ray enters a glass rod at a shallow angle. At point A the ray strikes the side of the rod, but because the angle θ is greater than the critical angle, total internal reflection occurs and the ray is contained within the rod. Similar reflections occur at points B, C and D until the ray emerges, unattenuated, from the end of the rod.

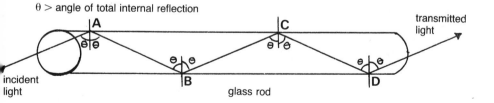

Fig. 8.21 Principle of fibre-optic transmission.

The principle of fig. 8.21 allows signals to be carried optically through glass or plastic rather than electrically down a wire. Figure 8.21 implies a rigid rod; in practice the use of many small diameter fibres as fig. 8.22a allows the construction of an optical 'cable' which is as flexible and strong as its electrical counterpart. A small data transmission system is shown in fig. 8.22b. The use of such 'cables' is known as fibre optics.

The use of fibre optics has several advantages. Noise in instrumentation is an electromagnetic phenomenon, so fibre optic transmission is totally free from noise interference, crosstalk or ground-loop problems and gives total electrical isolation between transmitter and receiver. Fibre optic cables can also be run with total safety through flammable and explosive atmospheres because sparking cannot result from cable breakage or damage.

Fibre optic cables also have wider bandwidth and lower transmission losses than coaxial cables when high frequencies are used. In electrical transmission the available bandwidth varies inversely as the square of the transmission distance; with fibre optics it varies inversely as the distance.

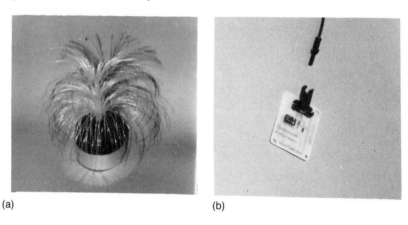

(a)   (b)

Fig. 8.22 (a) Fibre optic display. (b) Commercial fibre optic communication cable & termination.

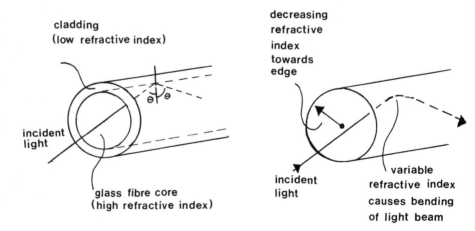

Fig. 8.23 Various constructions of fibre-optic cable. (a) Step index. (b) Graded index.

There are two types of fibre-optic construction, shown in fig. 8.23. *Step index* fibres (fig. 8.23a) are identical in principle to fig. 8.21, consisting of a central fibre of high refractive index surrounded by a thin cladding of low refractive index. Internal reflection takes place at junction between the fibre and the cladding.

Graded index fibre has a refractive index which varies gradually from a high value at the centre to a low value around the periphery. This causes the light rays to follow the gentler curves of fig. 8.23b. Graded index fibres have lower transmission losses and other advantages described below, but are obviously more expensive to manufacture.

The optical signal is attenuated as it passes down a fibre optic cable. This attenuation occurs from four main causes. The first two of these are directly related to the cable length and are quoted together by manufacturers as an attenuation in dB km^{-1}.

Material absorption occurs because impurities and manufacturing defects cause the optical signal to be gradually absorbed along the cable. Losses can be minimised by choosing an operating wavelength at which the losses are least. Fibre optic manufacturers quote a recommended operating region for their cable.

Scattering losses occur because of slight irregularities at the fibre/cladding interface, which means that a small proportion of the light rays strike the interface at an angle less than the critical angle and are absorbed by the cladding.

Losses also occur when the fibre follows a tight curve as this decreases the angle of incidence on the outer side of the cable as shown in fig. 8.24a. The curvature of the cable allows light which was previously outside the critical angle to be absorbed.

The final loss occurs at couplings between cable sections and between the cable and the transmitter and receiver. In fig. 8.24b the overall attenuation is given by:

$$\text{Attenuation} = 10 \log\left(\frac{P_T}{P_R}\right) \text{dB} \tag{8.9}$$

The attenuation consists of $(\alpha_T + \alpha_L + N\alpha_C + \alpha_R)$ where α_T and α_R are the transmitter and receiver to fibre losses (typically 3 dB for a plastic polymer system of 1 dB for a glass system); αL is the cable attenuation (typically 150 dB km^{-1} for polymer and 5 dB km^{-1} for glass) and $N\alpha_C$ is the attenuation for N in line connectors (each connector being similar to the

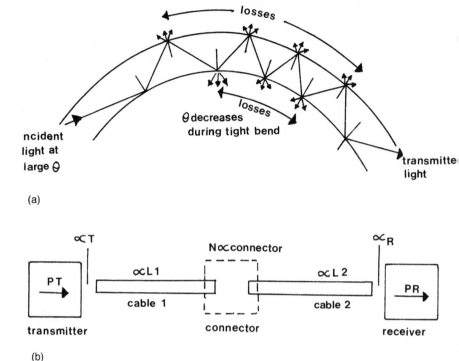

(a)

(b)

Fig. 8.24 Losses in fibre-optic transmission systems. (a) Losses caused by tight bend. (b) Transmission and coupling losses.

transmitter/receiver loss).

In fig. 8.25, a light beam is just being conveyed down the fibre and being reflected at just the critical angle each time. To do this, it must enter at angle θ_a which is known as the angle of acceptance. If n_a, n_f, and n_c are the index of refraction of air, the fibre and the cladding, analysis will show that:

$$\sin \theta_a = \frac{n_f}{n_a}\sqrt{1 - \left(\frac{n_c}{n_f}\right)^2} \qquad (8.10)$$

For all practical purposes, $n_a = 1$, so

$$\sin \theta_a = \sqrt{n_f^2 - n_c^2} \qquad (8.11)$$

The term 'numerical aperture' is often given to $\sin \theta_a$. Light rays entering at an angle greater than θ_a will not be conveyed down the cable. Typical values for the numerical aperture are 0.25 to 0.5, which gives angles of acceptance of between 15° and 30°.

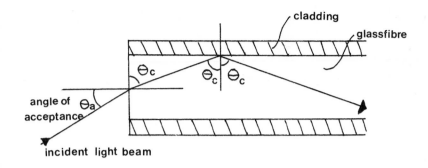

Fig. 8.25 Angle of acceptance.

The elements of a fibre optic system are therefore as shown on fig. 8.26. Data (in digital or some modulated form, e.g. AM, FM, PWM) is passed via a drive amplifier to an LED operating at a wavelength suitable for the cable (typically 600–850 nm). The light travels down the fibre optic cable and is received by a photodiode. The resulting diode current changes are amplified to give the original signal. A data transmission system similar to fig. 8.26 can have a bandwidth in excess of 100 MHz.

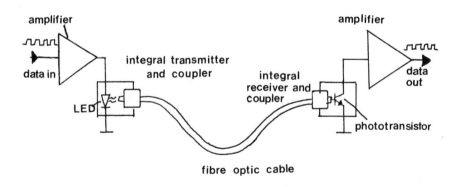

Fig. 8.26 Fibre-optic data transmission system.

8.7. Photocells

Photocells are devices used to detect the presence or absence of objects. They can be used, for example, to count objects passing down a conveyor belt or to replace a limit switch in some

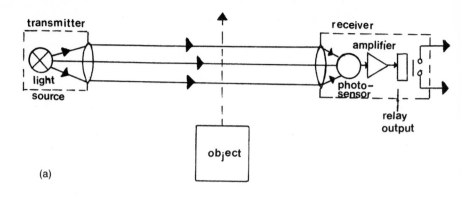

(a)

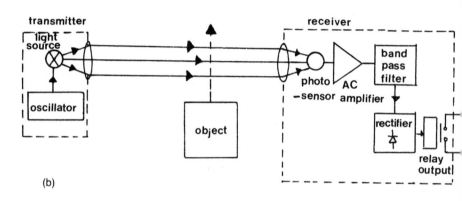

(b)

(c)

Fig. 8.27 Photocell systems. (a) Simple photocell transmitter and receiver. (b) Modulated photocell. (c) Photograph of photocell system.

circumstances. In essence the principle is very simple and is illustrated in fig. 8.27a.

A light source is placed at the focus of a lens to produce a parallel beam of light. This is received by another lens and focused on to a light sensor whose output drives a relay via a suitable amplifier. As objects break the light beam, the output relay de-energises.

The circuit of fig. 8.27a is, however, sensitive to all light, and not just that emerging from the photocell transmitter. It would, for example, probably respond to changes in ambient lighting. The circuit of fig. 8.27b is more commonly used and employs a modulated light source.

The source is driven from an oscillator at a high frequency (typically several tens of kilohertz) to produce a chopped light beam. In some photocells this chopping is done mechanically with a motor-spun mirror. The chopped light beam is received by a sensor whose output is an AC signal. This is amplified by an AC bandpass amplifier which will reject all frequencies other than the one used by the transmitter. The amplifier output is rectified and used to drive the output relay. Changes in ambient light produce a low frequency or DC change in the sensor output, and are correspondingly rejected by the bandpass amplifier. Temperature effects on the sensor are similarly ignored.

Figure 8.27 uses separate transmitters and receivers, both of which need a power supply. If this is inconvenient, a retro reflective unit can be used, one example of which is shown in fig. 8.28a. This incorporates the transmitter and receiver in one unit, and utilises a mirror to give an out-and-back light beam. The mirror is constructed of inverted prisms as shown in fig. 8.28b (similar to bicycle reflectors) which have the characteristic that any arriving beam is returned back along its path. Perfect alignment of the mirror is therefore not needed.

Retro reflective photocells can also be used to detect the presence of objects by using the object surface itself to backscatter the light. The range of such a detector is small, typically 100 to 300 mm depending on the object's reflectivity.

Fibre optics can be combined with photocells to advantage, allowing the 'delicate' electronics to be removed to less hazardous environments. Figure 8.28c shows a photocell that has a separate lens and detector head which focuses light on to the end of a fibre optic cable. The electronics transmitter and receiver circuits are mounted at the other end of the fibre optic cable. Apart from

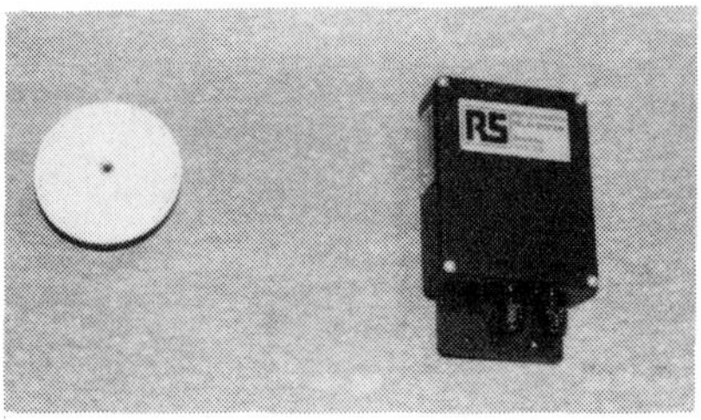

(a)

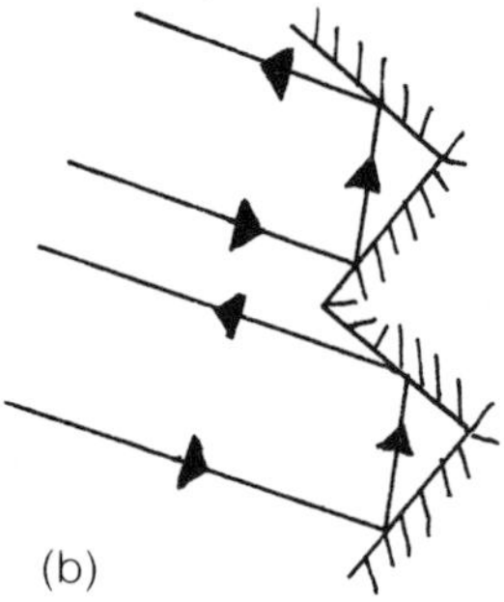

(b)

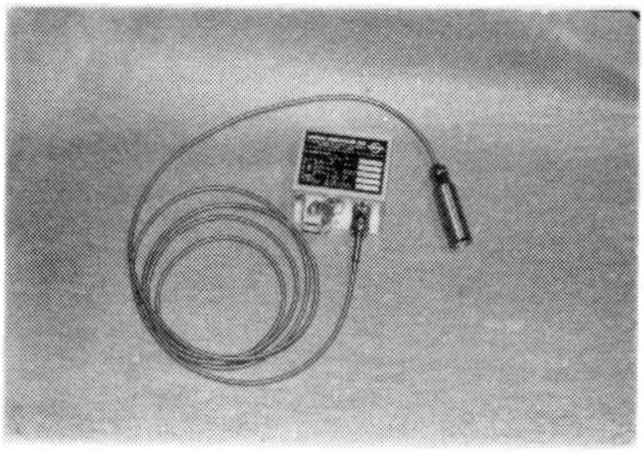

(c)

(d)

Fig. 8.28 Various photocells. (a) Retroflective unit & mirror. (b) Prismatic mirror used on retroreflective photocell. (c) Unit with fibre optic coupled head. (d) Infra red hot metal detector.

allowing the head to be mounted in, say, a high-temperature location while keeping the electronics cool, the fibre optic head assembly is much smaller than any integral unit.

8.8. Lasers

Light emission caused by the change of energy level of electrons in atoms was briefly described in section 8.5.4. Possible transitions for hydrogen atoms are shown in fig. 8.29. This shows that an energy input of at least 10 eV is needed to lift an atom to its first state; on falling back it emits light of frequency 1216 Å (ultraviolet). The light is emitted as a 'packet' of energy called a photon. Other transitions are also possible, with different wavelengths as shown. The larger the energy change, the higher the frequency. The relationship is given by:

$$E_2 - E_1 = hf \tag{8.12}$$

where h is Planck's constant. It follows that for a particular material emitting light via excitation from heat or electrical influences, there will be many specific wavelengths corresponding to all the possible transitions of fig. 8.29.

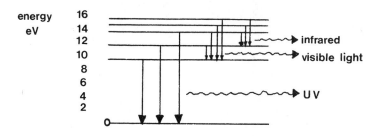

Fig. 8.29 Possible energy states for hydrogen atoms.

Laser light, however, is unique. It is monochromatic, i.e. it consists of light of just one wavelength. It is also coherent: all the photons emitted from the laser are exactly in phase. The difference between coherent and non-coherent light is shown in fig. 8.30. As a result of the coherence, monochromaticity and the mechanism by which it is produced, laser light is very intense and emerges as an exceptionally parallel beam. A typical divergence is less than 0.001 radian.

The construction of a typical ruby laser is shown in fig. 8.31a, and consists of a ruby rod with the ends machined parallel; one end is fully silvered to give an almost perfect mirror. The other is partially silvered. The ruby rod is surrounded by a flash tube.

Ruby has the energy states of fig. 8.31b. There is an energy state 1 at about 1.8 eV above the base state, and a large number (many hundreds) of energy states above state 1 which collectively

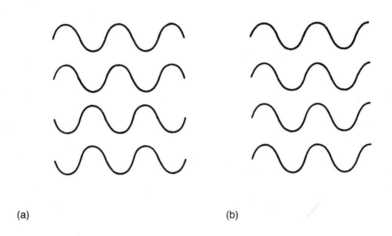

(a)   (b)

Fig. 8.30 Coherent and non-coherent light. (a) Single-frequency non-coherent light. Components out of phase, partial cancellation results. (b) Coherent light. All components are in phase and consequently reinforce each other.

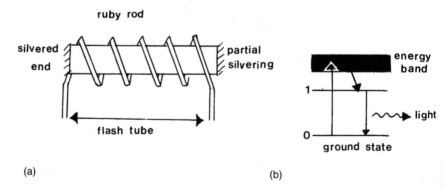

(a)   (b)

Fig. 8.31 A ruby pulsed laser. (a) Schematic construction. (b) Energy states and transitions of a ruby laser.

form an energy 'band'.

To start laser action, it is first necessary to get more atoms into the excited state than remain in the ground state. This is called a population inversion, and is achieved by 'pumping' energy into the rod by firing the flash tube for a short period. This takes the atoms in the ruby rod into the higher energy band of fig. 8.31b, from which they all spontaneously fall back to state 1 where they are transiently stable.

Eventually an atom will fall back to the base state, emitting a photon of light as it does so. If this photon strikes another excited atom (as is probable in a population inversion), the latter falls to the base state, emitting another photon which is exactly in coherence with the impinging photon. A positive feedback effect now takes place with a rapid build-up of photons, all of identical frequency and coherent. The silvered ends make the resultant rapidly intensifying light pulse traverse the rod several times before it emerges from the half-silvered end.

The transition from state 1 to the ground state is 1.8 eV, which by equation 8.12 corresponds to a wavelength of 6943 Å. This is in the red part of the spectrum, so an intense coherent red pulse of light lasting about a millisecond is produced.

The ruby laser of fig. 8.31 is called a pulse laser, for obvious reasons. For most industrial applications, a continuous beam is needed. This is not feasible with the ruby laser because the rod would not be able to absorb the energy necessary to maintain a permanent population inversion. To achieve this, a four-level laser system is used as shown in fig. 8.32.

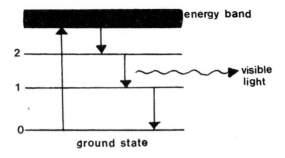

Fig. 8.32 Energy states for a continuous laser.

Atoms are pumped from the ground state to a high energy band from where they fall back to energy state 2. Laser action takes place between states 2 and 1, with atoms returning spontaneously to the ground state. To get laser action it is only necessary to maintain a population inversion between states 1 and 2, which is achievable with relatively low pumping energies. Figure 8.33 shows a typical industrial laser.

Fig. 8.33   Ferranti MFK–1200 Watt CO_2 laser cutting mild steel plate.

8.9.   Miscellaneous topics

8.9.1.   Spectroscopy

Light is emitted when excited atoms fall to a lower energy state, with the frequency being given by equation 8.12. The energy levels differ from element to element, so it is possible to identify the presence of elements, and their relative proportions, by observing the light emitted from a substance. Light from silicon, for example, contains a distinct spectral line at 2124 Å, and light from sodium contains two lines in the orange part of the spectrum (hence the colour of sodium street lights). These spectral lines can be used for chemical analysis (particularly in metallurgy). This analytical technique is called spectroscopy.

The basis of a spectroscopic analyser is shown in fig. 8.34a.

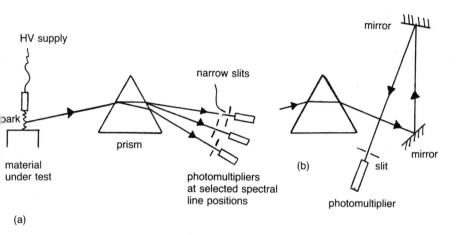

Fig. 8.34 Spectroscopical analysis. (a) Physical principle. (b) Increasing path length and resolution with folded beam.

Atoms of the substance to be analysed are excited, either by heating or by the striking of an electric arc. The emitted light passes through a prism and is split into its consitituent wavelengths.

A light detection device (usually a photomultiplier) is placed behind a narrow slit at a position corresponding to a spectral line of interest. The current from each photomultiplier indicates the relative amounts of the elements in the original substance.

Although the theory of fig. 8.34a is simple, there are many practical difficulties. Long path lengths on the exit side of the prism are necessary to give the requisite resolution. These are usually obtained by folding the exit beams around the inside of the instrument with mirrors, as in fig. 8.34b. Long path lengths, however, increase sensitivity to mechanical vibration so some form of cushion mounting is required.

Particular care needs to be taken to avoid errors due to contamination of the source light by absorption from the atmosphere around the prism. The source material must therefore be excited in an inert gas (e.g. argon), and the prism and photodetectors must be kept in a vacuum.

The final problem is that it is easy to detect the *presence* of an element, but difficult to infer qualitative results. Where specific proportions are needed (e.g. the copper percentage in a steel sample), comparisons with standard sources are used.

8.9.2. *Flame failure devices*

Flame failure detection is required for gas and oil burners to prevent the explosion hazard from an accumulation of unburnt fuel. For small burners a simple thermal detector will suffice, but in large boilers the thermal time constant is too long to shut the fuel off quickly.

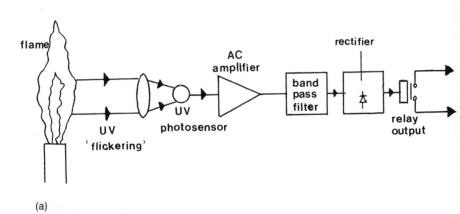

(a)

Fig. 8.35 Flame failure detector. (a) Principle of operation. (b) Unit monitoring oxy-gas burner.

Gas and oil flames emit ultraviolet light which 'flickers' at a predictable rate. This phenomenon allows an ultraviolet flame failure detector to be constructed as in fig. 8.35. Ultraviolet light from the flame is detected by a suitable detector whose output is amplified by an AC amplifier which passes the flicker but rejects the background UV. The amplifier output is rectified and used to energise an output relay.

Failure of the flame (or the detector head itself) will cause the output of the AC amplifier and rectifier to fall to zero and the relay to de-energise.

8.9.3. Photometry

Photometry is concerned with the measurement of light intensity, and as such is encountered in topics as diverse as photography and internal building illumination.

It is obvious that there is a difference between 'brightness' and the capability of a light source to illuminate an area. A 100-W tungsten filament bulb and a 60-W fluorescent tube have similar lighting capability, but the former is obviously brighter to the eye.

Lighting capability is defined as the luminous intensity and is measured in candelas (cd) approximating to the light from a standard candle. It being difficult to make a standard candle, the SI standard is based on the light emitted from a square centimetre of platinum at its melting point of 1773 °C, which has a luminous intensity of 60 cd.

Brightness is defined as the luminance, and is the luminous intensity divided by the area of the source. The SI source above, for example, has a luminance of 60 cd/cm². A fluorescent tube has a luminance of approximately 1 cd/cm² compared with 650 cd/cm² for a tungsten filament bulb. The much larger area of the fluorescent tube compared with the incandescent filament, however, gives a similar luminous intensity.

It is also possible to define the luminous flux, which relates the lighting capability of a light source to the area of the surface to be illuminated and the distance of the source. The light flux, ϕ, is measured in lumens, lm, and is defined as the light flux emitted in one steradian from a uniform candela source. (This can be envisaged as the light flux through an area of 1 m² at a distance of 1 m from a candela source: see fig. 8.36a.) By simple geometry, a given source of I candelas has a total luminous flux of $4\pi I$

lumens.

From the definition of light flux, it is possible to define the level of illumination at a given surface area (e.g. a desk top or work table). The level of illumination, in lux, is simply defined as the flux per square metre of surface, lm m^{-2} (obviously the area in fig. 8.36a has a level of illumination of 1 lx).

Suppose we have a 50-candela source (approximately a 60-W bulb). This has a total luminous flux of 628 lm. Assuming this is radiated equally in all directions, at a distance of 2 m, say, the flux will spread over an area of about 50 m^2 ($4\pi r^2$) giving a level of illumination of about 12.5 lx (628/50).

Because the area over which the luminous flux is dispersed increases as the square of the distance, it follows that luminous

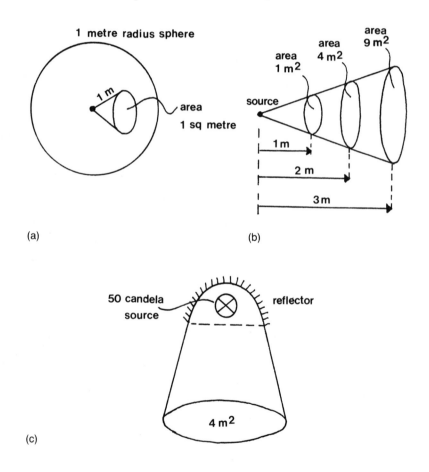

(a)

(b)

(c)

Fig. 8.36 Photometry principles. (a) Definition of level of illumination. (b) The inverse square law of illumination. (c) The effect of a reflector.

intensity falls off as the square of the distance, as illustrated by fig. 8.36b. A tripling of the distance gives one-ninth of the illumination.

Adding a reflector to a lamp can increase the level of illumination dramatically. In fig. 8.36c a reflector is used to focus the light from our 50-candela source on to a 4 m^2 area. The whole 628 lm are now spread over an area of 4 m^2, giving a level of illumination of 157 lx.

Light levels inside buildings are measured in lux. Recommended levels go from 100 lx in warehouses and general factory areas to about 750 lx for drawing offices. Light levels are generally measured by a light meter using a photovoltaic cell.

Chapter 9
Velocity, vibration and acceleration

9.1. Relationships

Velocity is the rate of change of distance, and acceleration is the rate of change of velocity. Velocity therefore has the units of m s^{-1} and acceleration m s^{-2}. Given any one quantity the others can be derived by integration or differentiation. (DC amplifier circuits are described in chapter 1, Volume 2.) Practical integrator circuits, however, suffer from long-term drift, and differentiators, by their nature, are very noise sensitive, so position, velocity and acceleration transducers are usually employed directly. Examples of non-direct position measurement are inertial navigation systems in spacecraft, where position in three axes is obtained from double integration of acceleration signals.

It is instructive to consider an object moving sinusoidally with respect to a fixed point (fig. 9.1). The position will be given by:

$$x(t) = K \sin \omega t \quad \text{m} \tag{9.1}$$

where K is the peak displacement and ω the angular frequency in rad s^{-1} (frequency in hertz is, of course, $\omega/2\pi$). The velocity is given by differentiation:

$$v(t) = K \omega \cos \omega t \quad \text{m s}^{-1} \tag{9.2}$$

and the acceleration by another differentiation:

$$a(t) = -K \omega^2 \sin \omega t \quad \text{m s}^{-2} \tag{9.3}$$

Velocity is at a maximum at the centre point of the motion, and acceleration at the extremes. Velocity and acceleration are both proportional to K, the amplitude. Note, however, that velocity is proportional to the frequency, and acceleration proportional to

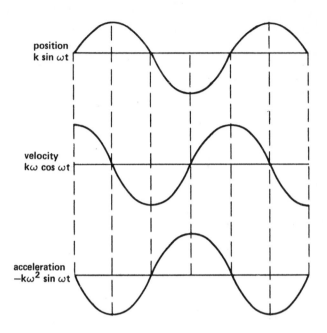

Fig. 9.1 The relationship between position, velocity and acceleration.

the square of the frequency.

It follows that very large accelerations can occur with small displacements. It is, in fact, the acceleration required by the needle that limits the frequency response of a hi-fi unit. Similarly, small-amplitude high-frequency vibration of mechanical plant can cause damaging accelerations and forces. The measurement of vibration is essential in large rotating plant such as fans, compressors, etc.

Acceleration is often measured with respect to the acceleration due to gravity (9.8 m s^{-2}). An acceleration of 29.4 m s^{-2} can, for example, be represented as $3\ g$. This notation is generally used for large transient acceleration.

9.2. Velocity measurement

9.2.1. Tachogenerator

Many applications require measurement of angular velocity (e.g. shaft speeds of motor-driven plant) and in many others it is convenient to convert a linear movement to an angular movement (e.g. by a rack-and-pinion) for measurement purposes.

The commonest, and simplest, angular velocity transducer is

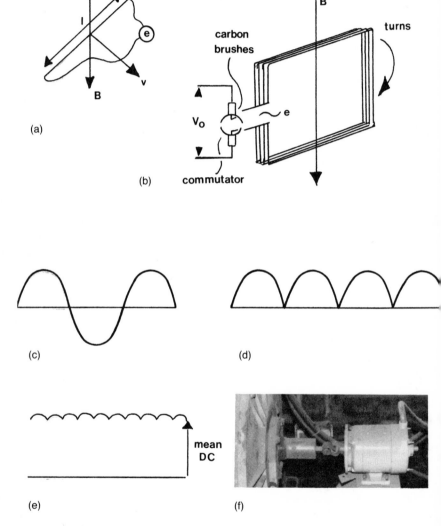

Fig. 9.2 The tachogenerator. (a) Wire moving in a magnetic field. (b) Coil rotating in a magnetic field. (c) Induced coil voltage, *e*. (d) Output voltage; converted to DC by commutator. (e) Output from multiple-commutator segment generator. (f) Photograph of tachogenerator.

the tachogenerator. This is effectively a simple DC generator. The principle is discussed further in chapter 2, Volume 2, but is shown here in fig. 9.2. A wire, length l, is moving with velocity v perpendicular to a magnetic field of flux density B. A voltage e is induced in the wire where:

$$e = B\,l\,v \tag{9.4}$$

In fig. 9.2, a coil of n turns is rotating in a field B. Each side of each turn parallel to the axis of rotation will have a voltage induced as above. The turns are not, however, moving perpendicular to B. If the coil is at an angle θ, the induced voltage is:

$$e = B\,l\,v \sin \theta t \tag{9.5}$$

But $v = r\omega$ where r is the coil radius, and ω the angular velocity (in rad s^{-1}). The coil angle, θ, also varies with time as ωt where t is in seconds. Both halves of the coil have an identical voltage induced, so the net coil output is:

$$e = 2\,B\,l\,r\,\omega\,n \sin \omega t \tag{9.6}$$

This is a sinusoid whose frequency and peak amplitude are proportional to the angular frequency, ω, as shown in fig. 9.2c.

An AC output tachogenerator is used, and the coil voltage can be measured via slip rings. It is usually more convenient, though, to have a DC output. The commutator segments on fig. 9.2b reverse the coil connections as the coil voltage passes through zero to give a rectified DC output voltage as in fig. 9.2d. The DC level is proportional to angular velocity, but the ripple content is unacceptably large and, worse, the ripple frequency is speed dependent.

The ripple can be reduced to acceptable levels by using more coils and commutator segments to give an output voltage similar to fig. 9.2e. Ten to twenty commutator segments are common; the larger the number the less the ripple, but the more complex and expensive the device. Ripple will be about 1–2% of the output voltage in a well-designed tachogenerator, and can be reduced by subsequent filtering if the inherent filtering lag can be tolerated. Ripple frequency will be 2.N.shaft frequency where N is the number of coils.

Tachogenerators are not designed to supply a large current, and should be connected to a high impedance load (typically 100 k-ohm). They are usually calibrated in terms of revolutions per minute; a common standard is 10 V per 1000 rpm. Speeds up to 10 000 rpm can be measured directly, the limiting factor being centrifugal force on the commutator segments.

9.2.2. Drag cup

The drag cup converts angular velocity to angular displacement, and is commonly found in motor-car speedometers. If remote indication is required, the angular displacement can be converted to an electrical signal by any of the angular displacement transducers in chapter 4.

The principle of the drag cup is shown in fig. 9.3. A concentric cylindrical magnet and keeper cup are rotated by the input shaft. The magnet has four poles, giving flux lines as shown in the cross-sectional insert. The drag cup itself is a conducting, non-ferromagnetic (e.g. copper) cup connected to the output

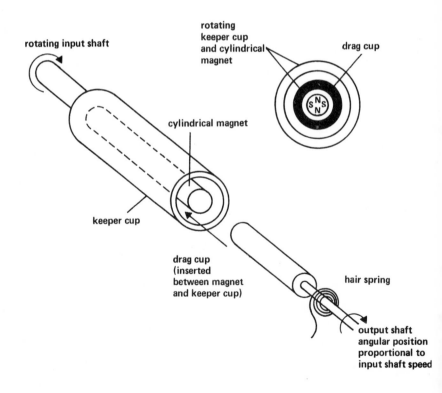

Fig. 9.3 The drag cup angular velocity transducer.

shaft and fitted, again concentrically, between the magnet and the keeper.

As the input shaft rotates, the flux lines between the magnet and keeper move with respect to the drag cup. This induces voltages as in equation 9.4, and induced currents flow in the drag cup. These currents, reacting with the magnetic field, produce a torque on the cup which, in the absence of any restraining torque, would rotate in the same direction as the input shaft. (The action is very similar in principle to the AC induction motor described in chapter 2 in Volume 2.)

The output shaft is restrained by a torsional spring which provides a torque proportional to shaft rotation. The shaft assumes an angle at which the torque from the spring balances the torque induced by the rotating magnet. The shaft angle is then proportional to input shaft angular speed.

9.2.3. Pulse tachometers

A pulse tacho is a device which produces a constant amplitude pulse train output whose frequency is related to input shaft speed. They are usually specified by the number of pulses per revolution of the input shaft. A device, for example, with 180 pulses per revolution would have an output frequency of 9 kHz at 3000 rpm.

There are two common ways of converting the pulse train to a signal for control or display. Figure 9.4a uses a monostable with constant width output pulse which is edge triggered by the tacho pulses. The monostable output is passed through a low-pass filter to give a DC output proportional to speed. The maximum speed is determined by the monostable pulse width.

An alternative technique, shown in fig. 9.4b, directly counts the tacho pulses over a fixed period of time. (As such it is, of course, a sampling system whose sampling rate must be chosen with due consideration for the dynamics of the rest of the equipment being controlled or monitored.) The counter output can be transferred directly to a display (as in fig. 9.4b) or converted to an electrical signal via a digital-to-analog converter (DAC). Digital counters and DACs are discussed in chapter 3 in Volume 2.

One interesting application of pulse tachos is in the area of DC motor speed control. Phase locked loop (PLL) ICs generate an output voltage proportional to the frequency diffference between

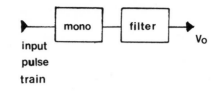

input
pulse
train

(a)

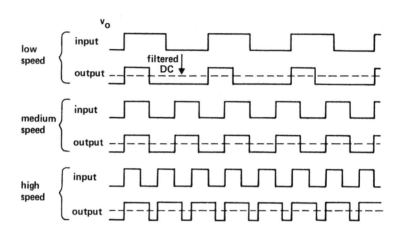

(b)

Fig. 9.4 Pulse tacho circuits. (a) Monostable/filter circuit. (b) Timed counter circuit.

two pulse trains. In fig. 9.5 the set speed is converted to a pulse train by a voltage-controlled oscillator (VCO) and compared with the output of a pulse tacho by a PLL. The output from the PLL is used as feedback for a P + I controller and power amplifier. Using the technique of fig. 9.5, excellent speed accuracy (better than 0.01%) can be achieved.

Most pulse tachos use optical techniques and are identical to

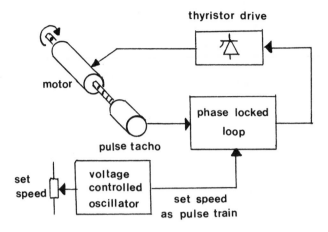

Fig. 9.5   Pulse tacho–based speed control.

the optical incremental position encoders described in section 4.4.2. In many numerically controlled machine tools, in fact, the same device serves as both position and velocity transducer.

Ad hoc optical pulse tachos can also be made by attaching reflective tape to a shaft and detecting reflections as in fig. 9.6. Commercially available devices based on fig. 9.6 allow non-contacting speed measurement with hand-held instruments.

Magnetic variable reluctance transducers can also be used with

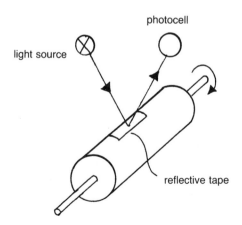

Fig. 9.6   'Ad hoc' pulse tacho using photoreflective method.

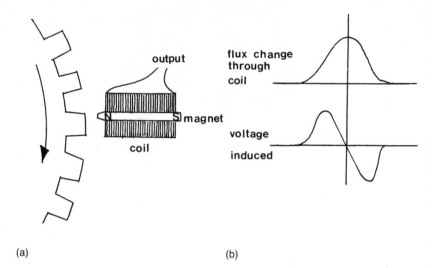

output

flux change
through
coil

magnet

coil

voltage
induced

(a)   (b)

Fig. 9.7 Variable reluctance tachogenerator. (a) Physical arrangement. (b) Waveforms.

a toothed wheel, as shown in fig. 9.7. As the teeth pass the magnet, the flux through the coil changes, causing an output voltage as shown in fig. 9.7b. One disadvantage with the variable reluctance tacho is the fact that its output voltage amplitude is also speed dependent. Additional electronics (e.g. Schmitt triggers) are required to give a constant voltage pulse train. The variable level of output voltage also sets a minimum speed.

9.2.4. Tachometer mounting

The coupling of a tachometer to a shaft is very critical. Vertical or horizontal misalignment can produce side forces on the tachometer bearings and early failure. Angular misalignment can cause a cyclic error between the input and the tachometer shaft. Above all there must be no backlash in the coupling, particularly in speed control systems. Backlash in a closed loop speed control appears as vicious speed hunting which damages motors, couplings and gearboxes.

Tachos are usually coupled by hysteresis-less universal joints similar to fig. 9.8 or (sideways displaced) by toothed belts. Tension adjustment on belts is critical to avoid bearing wear. Axial mounted tachos require vertical or horizontal adjustment;

Fig. 9.8 Flexible coupling linking motor and tachogenerator. The tacho is a pulse tacho and the unit before the coupling is a mechanical overspeed switch.

often this is achieved with shims.

9.2.5. Doppler systems

Doppler shift occurs when there is relative motion between a source of sound, or electromagnetic radiation (radio, radar or light). It is commonly observed as a change in pitch of a car horn or train whistle as the vehicle passes.

In fig. 9.9a, the source S is emitting waves (sound or electromagnetic radiation) with a frequency f. If the velocity of propagation is c, and the wavelength λ, these are related by:

$$f = c/\lambda \tag{9.7}$$

The stationary observer will see f wavefronts pass per unit time, i.e. the observed frequency is the same as the transmitted frequency.

If the observer is now moving towards the source with velocity v, as in fig. 9.9b, each wavefront will be observed earlier, as the apparent velocity of propagation is $c + v$. The observed frequency is:

$$fv = (c + v)/\lambda \tag{9.8}$$

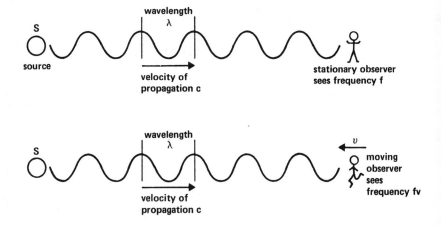

Fig. 9.9 The Doppler effect. (a) Stationary observer. (b) Moving observer.

i.e. a higher frequency. The frequency shift is:

$$fv - f = \frac{c + v}{\lambda} - \frac{c}{\lambda} \tag{9.9}$$

$$= v/\lambda \tag{9.10}$$

$$= fv/c \tag{9.11}$$

Equations 9.10 and 9.11 show that the frequency shift is proportional to the relative velocity and the original frequency. A similar result is obtained for a receding observer, except that the frequency is lowered. Identical results are obtained for a fixed observer and a moving source since the Doppler effect arises from *relative* motion.

The Doppler shift allows remote measurement of velocity (it has, for example, been used to measure the rotational velocity of planets by observing the shift in frequency of radar pulses reflected from opposite sides). The principle is shown in fig. 9.10. A transmitter (ultrasonic, radar or light) emits a frequency f_t at velocity c. The object, moving at velocity v, reflects part of this which is Doppler-shifted to f_r. Two Doppler shifts occur as the object is acting both as a moving observer and as a moving source, so

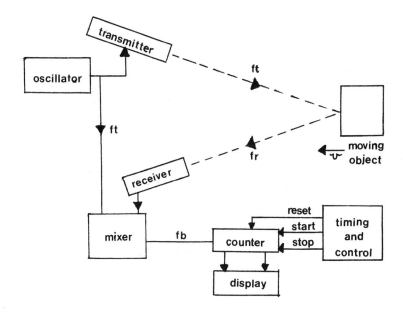

Fig. 9.10 Doppler shift velocity measurement.

$$f_r - f_t = \frac{2f_t}{c} v \qquad (9.12)$$

The transmitted and received frequencies are mixed to give a beat frequency f_b equal to the shift $f_r - f_t$. The beat frequency is proportional to v, and can be converted to a display or signal by a counter or frequency-sensitive rectifier.

It is instructive to observe the magnitude of the effect, which is dependent on the ratio v/c. For a radar signal of 10 GHz (10^{10} Hz) travelling at the speed of light 3×10^8 m s^{-1} and an object moving at 20 m s^{-1}, the shift is 1.333 kHz (i.e. an audio frequency).

9.3. Accelerometers

9.3.1. Seismic mass accelerometers

If a body of mass M kg experiences a force F newtons, it will accelerate at a m s^{-2}, these three quantities being related by Newton's second law of motion:

$$F = Ma \tag{9.13}$$

It also follows that a body which is accelerating will experience a force; we feel a force that pushes us into our seats when a car accelerates, and a force forwards into the seat belt when it brakes. This force is proportional to the acceleration and can be used to measure the acceleration.

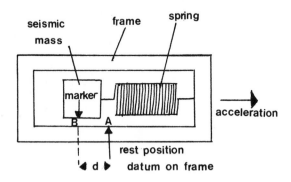

Fig. 9.11   Seismic mass accelerometer.

The principle is shown in fig. 9.11. A mass (called a seismic mass) is restrained by a spring of stiffness K. In the rest position, the mass is at position A. If the body accelerates at a constant rate a, the mass experiences a force according to equation 9.13. This causes the mass to move to the left to position B, until the force caused by acceleration balances the restoring force of the spring. In balance:

$$Ma = dK \tag{9.14}$$

i.e. the displacement is proportional to the acceleration.

The displacement can be measured by any displacement transducer: strain gauges, potentiometers and LVDTs are common as we shall see in section 9.3.3.

9.3.2.   Second-order systems

Figure 9.11 is essentially similar to the suspended mass of fig. 9.12 with acceleration and gravity fulfilling the role of a displacing force. If the mass in fig. 9.12 is pulled down and released, we would, by experience, expect it to oscillate with decreasing amplitude until it returns to its original position. It is reasonable to expect similar oscillations from the seismic mass in an accelerometer.

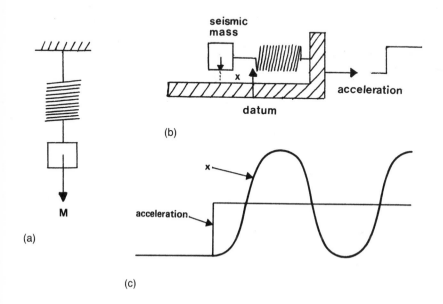

Fig. **9.12** Step response of undamped accelerometer. (a) Oscillating spring/mass. (b) Seismic mass accelerometer with step change of acceleration. (c) Response of accelerometer.

In fig. 9.12b, a seismic mass accelerometer is used to measure a step change in acceleration, a. The mass will lag behind the frame, giving a net mass acceleration of $[a - (d^2x/dt^2)]$.

Equation 9.14 can be rewritten:

$$M\left(a - \frac{d^2x}{dt^2}\right) = xK \tag{9.15}$$

or

$$\frac{d^2x}{dt^2} + \frac{Kx}{M} = a \tag{9.16}$$

If we denote K/M by ω_n^2 (for reasons which will become apparent), equation 9.16 becomes:

$$\frac{d^2x}{dt^2} + \omega_n^2 x = a \qquad (9.17)$$

the solution of this second-order equation is a sinusoid of the form:

$$x = \frac{Ma}{K} (1 - \cos \omega_n t) \qquad (9.18)$$

which is shown in fig. 9.12c. The theory predicts constant, non-decaying sinusoid of frequency $\omega_n/2\pi$ (called the natural frequency). Note that ω_n is $\sqrt{(K/M)}$, and the mean value is given by the steady state displacement Ma/K.

In practice, of course, friction will cause the oscillations to decay, but obviously the seismic mass/spring accelerometer cannot be used in the simple form of fig. 9.11. Some form of controlled, predictable damping is needed. Simple constant frictional force will damp the oscillation but give an offset error. A viscous force, proportional to velocity, will damp the oscillation and cause no error. This characteristic can be obtained from a liquid dashpot (and is commonly found as shock absorbers on motor-car suspensions where they are used to damp out oscillations from the car springs).

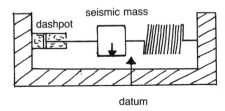

Fig. 9.13 Accelerometer with damping.

The revised arrangement is shown in fig. 9.13. The opposing force now is

$$Kx + C\frac{dx}{dt}$$

so equation 9.15 becomes:

$$M\left(a - \frac{d^2x}{dt^2}\right) = Kx + C\frac{dx}{dt} \qquad (9.19)$$

which can be rearranged:

$$\frac{d^2x}{dt^2} + \frac{C}{M}\frac{dx}{dt} + \frac{K}{M}x = a \qquad (9.20)$$

The solution of this equation has two parts, a transient part plus a steady state part (as t tends towards infinity). The steady state part is simply $x = Ma/K$ as before. To examine the transient part we define a damping factor b, where:

$$b = \frac{C}{2\sqrt{mK}} \qquad (9.21)$$

Defining ω_n as before, equation 9.20 becomes:

$$\frac{d^2x}{dt^2} + 2b\omega_n\frac{dx}{dt} + \omega_n^2x = a \qquad (9.22)$$

The mathematical analysis of this equation is somewhat lengthy and beyond the scope of this volume, but the predicted result depends on the value of b.

For $b<1$ the system will exhibit oscillations that decay exponentially as shown in fig. 9.14a. The system is said to be underdamped. The case $b = 0$ corresponds to no damping and to equation 9.17 and fig. 9.12c.

For $b>1$ the system does not oscillate, and rises as in fig. 9.14c. The system is said to be overdamped. The case of $b = 1$ marks the transition from overdamping to underdamping, as in fig. 9.14b, and is called critical damping.

Figure 9.15 shows superimposed step responses for various damping factors. Note that the curves are normalised with respect

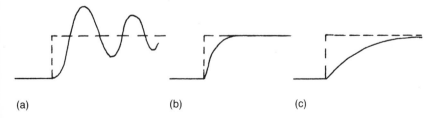

(a)   (b)   (c)

Fig. 9.14 The effect of damping. (a) Underdamped ($b<1$). (b) Critically damped ($b = 1$). (c) Overdamped ($b>1$).

to ω_n. Figure 9.15 also shows that the actual oscillation period increases from ω_n with increasing damping.

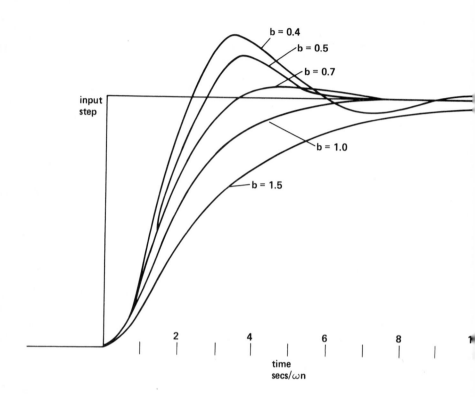

Fig. 9.15   Step response for various damping factors.

It might be thought that the best response occurs for a damping factor $b = 1$, but in practice slight underdamping is preferred. Examination of fig. 9.15 shows that the shortest time to settle within any specified error band occurs for $b = 0.7$ (corresponding to a first overshoot of about 8%).

Figure 9.16 shows the Bode diagram for various values of damping factor. Again a damping factor of 0.7 gives the largest usable bandwidth. The Bode diagram also demonstrates the importance of knowing the frequency range of the acceleration being investigated. In general a second-order transducer is usable up to one-fifth of its resonant frequency.

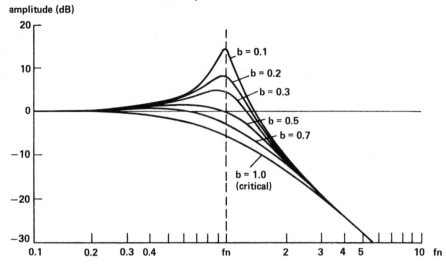

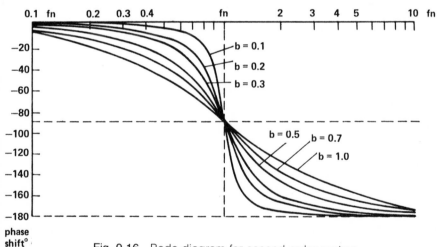

Fig. 9.16 Bode diagram for second-order system.

9.3.3. Practical accelerometers

The transducer in fig. 9.17 uses a seismic mass suspended from thin flexure plates. The mass is restrained by thin wires which also act as strain gauges for deflection measurement. As the device accelerates to the right, for example, gauges A and B will decrease in length, and C and D increase in length. These can be connected into a bridge circuit as described in section 6.3, to give a voltage output that represents acceleration. The use of four gauges gives temperature compensation and an increased output signal. The spring is provided by the flexure plates and the strain gauges themselves. Viscous damping is obtained by filling the transducer with oil; the motion of the flexure plates and the mass itself giving the required damping force.

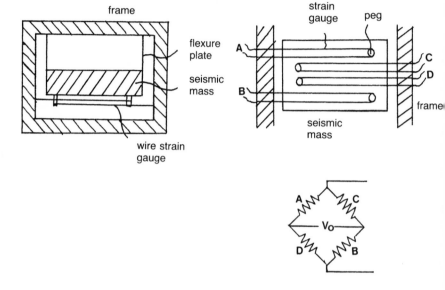

Fig. 9.17 Strain gauge accelerometer. (a) Side view. (b) Top view. (c) Gauge connections in bridge.

An alternative approach is the swinging mass of fig. 9.18. For simplicity of drawing, the support frames and bearing have been omitted. The sensing axis runs perpendicular to the line joining the two shafts. As the device accelerates, the two masses rotate, the restoring force being provided by springs. The acceleration is measured by an angular position transducer on one shaft, and damping provided by a dashpot on the other.

Accelerometers must only respond to acceleration along one axis, and ignore acceleration in other directions. For complete measurement of acceleration in three dimensions, three transducers are needed. The ability of an accelerometer to ignore non-axial acceleration is defined by its cross-axis sensitivity (also called the transverse sensitivity). It is usually defined as the ratio of the output of the device for acceleration perpendicular to, and

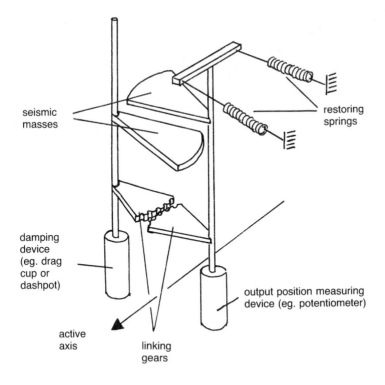

seismic
masses

restoring
springs

damping
device
(eg. drag
cup or
dashpot)

output position measuring
device (eg. potentiometer)

active
axis

linking
gears

Fig. 9.18   Swinging mass accelerometer.

along, the sensing axis. Typical values are around 2%. In fig. 9.18 cross-axis sensitivity is reduced by the use of two seismic masses, and in fig. 9.17 the flexure plates only permit movement along the sensing axis.

Figure 9.19 uses a force balance technique, matching the *Ma* force with an electromagnetic force. As the transducer accelerates to the right, the mass lags and moves with respect to the housing. The deflection is measured by an LVDT, and an amplifier increases the current in the restoring coil to bring the mass back to its rest position. The coil current is then a measure of the acceleration.

The piezo-electric effect, described in section 3.4, can also be used to measure acceleration. A piezo-electric crystal is essentially a force-measuring device, and as such can be used to measure the *Ma* force directly. The principle is shown in fig. 9.20. The seismic mass is in direct contact with the crystal, the latter being kept in permanent compression by the pre-tensioning

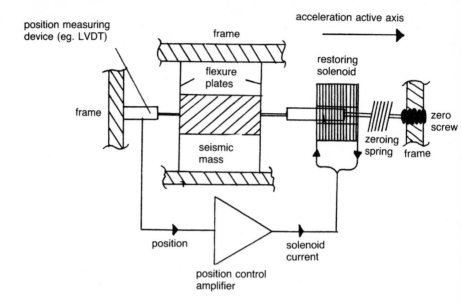

Fig. 9.19 Force balance accelerometer.

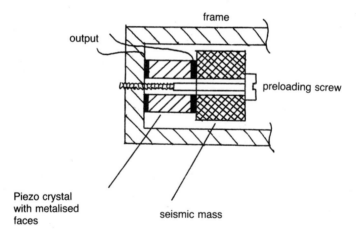

Fig. 9.20 Piezo-electric accelerometer.

screw. Acceleration along the sensing axis produces an *Ma* force on the crystal, giving an output as described in section 3.4.

The piezo-electric transducer requires a charge amplifier (also described in section 3.4) and has poor low-frequency response. It is best suited for measurement of high-frequency, high-*g*

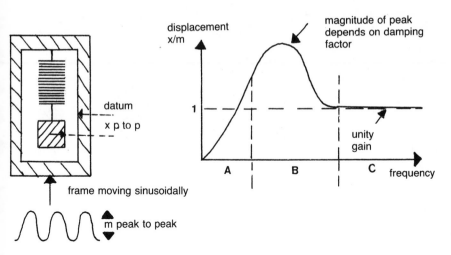

Fig. 9.21 Principle of vibration transducers. (a) Schematic diagram. (b) Frequency response.

acceleration, as found in impact testing. Piezo-electric transducers have the advantage of robustness, simplicity and small size.

9.4. Vibration transducers

Consider the seismic mass transducer of fig. 9.21a, which is being driven by a sinusoidal displacement of constant amplitude but variable frequency. The relationship between the displacement x and the applied frequency is shown in fig. 9.21b

In the region A, the device is acting as an accelerometer, the amplitude increasing because the acceleration increases with frequency. Region A extends to up to about one-fifth of the resonant frequency. As the frequency increases, there is a resonant region B after which the mass ceases to move in space, remaining fixed while the frame moves about it.

This corresponds to region C where the amplitude of the apparent displacement x follows the frame movement (but displaced by 180°). This region starts at about five times the resonant frequency, and extends, in theory, to infinite frequency.

A vibration transducer operating in region C becomes a

displacement transducer, and as such can be used to measure high-frequency vibration if a position transducer is attached between the seismic mass and the frame. If a velocity transducer is used, vibration velocity can be measured.

Vibration transducers are widely used for fault monitoring and protection of large rotating plant. Vibration transducers must inherently be small and of low mass to avoid loading the device to which they are attached. Like accelerometers they are second-order devices, and are usually designed for a damping factor of 0.7.

Index

absolute position transducer, 84, 107
absolute pressure, 63
absolute zero, 20
absorbtion (of radiation), 207
acceleration, 256
accelerometers, 267
accuracy, 9
ageing effects, 12
ALARA, 212
alpha radiation, 212
angle of acceptance, 242
angle of incidence, 221
angle of refraction, 221
Annubar, 138
anticipation, 187
apex balancing, 176
atmosphere, 63
atomic light emitters, 238
attenuation (of radiation), 207
axial strain, 174

backlash (in gears), 11
bar, 63
batch weighing, 187
belt weighers, 189
bending strain, 174
bequerel, 212
beta radiation, 212
bimetallic thermometers, 23
black body radiation, 52
bode diagram, 19
bolometer, 55
Bourdon tube, 67
bridge amplifiers, 177
bridge balancing, 176
bridge circuits, 170
British Standards,
    BS1042, 130
    BS1094, 28

BS1843, 45
BS3680, 158
bulk solids, level measurement, 196, 204

candelas, 253
capacitive sensing element, 70, 202
CDX, 104
Celsius scale, 20
centigrade scale, 20
characteristic temperature, 33
charge amplifier, 74
Charles's law, 23
chromatic aberration, 225
Cipoletti weir, 158
coarse/fine system, 101
coding (of shaft encoders), 107
coherence, 247
cold junction compensation, 45
colour coding of thermocouple cables, 46
compensating cable, 42
compressible fluids (flow of), 130
concave,
    lens, 224
    mirror, 219
constant current source, 30
control synchro, 91
control transformer, 98
control transmitter, 101
convex,
    lens, 223
    mirror, 219
corner tapping, 134
Cowan circuit, 100
critical angle, 221
cross correlation flowmeter, 153
cross sensitivity,
    of accelerometers, 274

of strain gauges, 167
CT, 104
curie, 212
current loop, 3
current-to-voltage conversion, 5
CX, 104

Dall tube, 133
damping factor, 17, 271
dark current, 230
definitions, 6
Delta P transmitters, 62
diaphragm sensing element, 70
differential amplifier, 178
differential pressure, 62
differential pressure flowmeters, 126
differential synchros, 101
disappearing filament pyrometer, 59
discharge coefficient, 129
displacement transducers, 116
Doppler flowmeter, 150
Doppler velocity measurement, 265
dose equivalent, 212
drag cup, 260
drift, 12
dynamic effects, 14
dynamic pressure, 62
dynamometer, 181
dynode, 231

E thermocouples, 40
E transformer, 118
elastic limit, 162
elastic modulus, 163
elastic sensing elements, 67
electromagnetic flowmeter, 148
electromagnetic radiation, 215
electromagnetic spectrum, 51
emissivity, 52
encoder (shaft), 107
energy bands, 238, 247
environmental effects, 12
error, 9
error band, 13
expansion thermometers, 21
exponential lag, 16
extension cable, 42

Fahrenheit scale, 20
feeder systems, 187
fibre optics,
    fibre types, 241
    losses in, 241
    principles, 239

first order systems, 14
flame failure devices, 232
flange tapping, 134
floats, 193
flowmeters,
    cross correlation, 153
    definition of flow, 125
    differential pressure, 126
    electromagnetic, 148
    flow energy balance, 127
    hot-wire anemometer, 156
    injection, 157
    open channel, 158
    orifice plate, 131
    Pitot tube, 137
    turbine, 142
    ultrasonic, 150
    variable area, 144
    venturi, 133
    vortex shedding, 145
focus and focal length,
    lens, 223
    mirror, 220
foil strain gauges, 167
force balance pressure transducer,
    electrical, 76
    pneumatic, 77
force balance systems, 76
force balance weigher, 159
frequency response, 17
FSD, 9
full scale deflection, 9
fundamental interval, 27

gamma radiation, 208, 212
gas pressure thermometers, 23
gas reaction (level measurement), 200
gauge factor, 166
gauge pressure, 63
Geiger Müller Tube, 209
GM tube, 209
graded index fibre, 241
gray (radiation unit), 212
gray code, 111
grounded junction thermocouple, 42

half life, 208
head loss, 129
head pressure, 63, 65, 197
hot-wire anemometer, 156
hysteresis, 11

ice-cell, 45
illumination, 253

in-flight compensation, 187
incandescent emitters, 237
inclined tube manometer, 66
incompressible fluid (flow of), 128
incremental encoder directional
    output, 115
incremental position transducer, 84,
    115
inductive potentiometer, 106
infrared, 53, 216
injection flow measurement, 157
instrumentation system, 1
insulated junction thermocouple, 42
intermediate metals (law of), 38
intermediate temperatures (law of),
    38
internal reflection, 221
invar, 23
ionisation gauge, 83

J thermocouple, 40

K thermocouple, 40
Kelvin scale, 20
kinetic energy (of a liquid), 127
Kirchhoff's law, 54

laminar flow, 127
LCD, 235
lead resistance,
    effect of bridge circuits, 176
    effect on RTDs, 32
lead transposition (of synchros), 95
LEDs, 233
lenses, 221
level measurement,
    capacitive probe, 202
    floats, 193
    gas reaction, 200
    general, 191
    head pressure, 197
    nucleonic, 207
    of solids, 196, 204
    resistive probe, 204
    resonance, 193
    switches, 213
    ultrasonics, 204
light emitters, 233
light emitting diodes, 233
linear devices, 7
linear variable differential transducer,
    71, 116
linearisation, 9
liquid crystal displays, 235

loadcells,
    construction, 184
    general, 183
    mounting, 186
    protection, 187
loading effect (on potentiometer), 88
lumens, 253
luminance, 253
lux, 254
LVDT, 71, 116

magnetoelastic effect, 181
magslip, 95
manifolds, 140
manometer, 65
mass attenuation coefficient, 207
mass flow, 125
measurand, 1
microstrain, 161
mirrors, 217
modulus of elasticity, 163
Moiré fringe, 113
motion detection, 187
mounting,
    of Delta P transmitters, 140
    of loadcells, 186
    of pressure transmitters, 79
    of strain gauges, 169
    of tachometers, 264

natural frequency, 17
negative temperature coefficient
    thermistor, 35
neon lamp, 238
Nixie tube, 239
noise immunity, 3
normalisation (of flow), 125
nozzle (flow measurement), 133
NTC thermistors, 35
nucleonic transducers, 207
null balance weigher, 159
numerical aperture, 242

offset zero, 6
open channel flowmeters, 158
optics, 217
optoelectronics,
    electromagnetic radiation, 215
    emitters, 233
    fibre optics, 239
    lasers, 247
    photocells, 243
    reflection, 217
    refraction, 221

sensors, 227
spectrum, 216
orifice plates,
general, 131
identification, 131
tappings, 134
types, 134
overpressure protection, 78
overshoot, 17

parallel balancing, 176
pascal, 63
phase locked loop, 261
phase sensitive rectifier, 100
phosphorescence, 238
photocells, 243
photodiodes, 229
photometry, 253
photomultiplier, 231
photoresistors, 227
phototransistors, 230
photovoltaic cells, 231
piezo-electric accelerometer, 275
piezo-electric effect, 73
piezo-resistive transducers, 75
Pirani gauge, 82
Pitot tube, 137
plane of vena contracta, 129
Planck's constant, 247
plate tappings, 134
plateau (of GM tube), 210
platinum resistance thermometers, 28
PLL, 261
Poisson's ratio, 164
position measurement,
general, 84
potentiometers, 85
resolvers, 105
shaft encoders, 107
synchros, 90
potential energy (of a liquid), 127
potentiometers, 85
preact, 187
pressductor, 183
pressure,
absolute, 63
differential, 62
dynamic, 62
gauge, 62
head, 63, 197
static, 62
units, 63
pressure transducers,
installation of, 79

practical notes, 79
specification of, 78
types, of, 62
primary sensor, 2
prism, 225
process variable, 1, 6
proof pressure, 78
proof ring, 183
proving frame, 183
proximity detectors, 122
PT100, 28
PTC thermistors, 35
pulse tacho, 261
pumping (of a laser), 249
PV, 6
pyrometery,
instruments, 58
principles, 55
theory, 51

quality factor, 212

R thermocouple, 40
rad, 212
radiation pyrometry, 51
radioactive sources legislation, 211
radius tap, 133
range, 7
range tubes, 66
RC, 104
receiver (synchros), 94
real image, 221
rectifier (phase sensitive), 100
reflected code, 111
reflection, 217
reflective mode (of LCD), 236
refraction, 221
refractive index, 221
rem, 212
repeatability, 11
resolution, 10
resolvers, 105
resonance (of liquids), 193
resistance thermometers,
connection of, 28
principles, 26
retroflective photocell, 245
Reynolds number, 127
rise time, 16
rotor (of synchros), 92
RTD, 27
ruby laser, 248
RX, 104

S thermocouple, 40
scale factor, 7
scintillation counter, 210
SDC, 127
sealed pressure gauge, 63
second order systems, 17, 268
Seebeck effect, 36
seismic mass, 267
Selsyn, 91
sensitivity, 7
sensor, 1
sensor (primary), 2
shaft encoders, 107
shear modulus, 179
shear strain, 164
signal,
   3–15 psi, 6
   4–20 mA, 3
   standards, 3
Snell's Law, 221
span, 7
spectroscopy, 250
spectrum, 51, 216
square root response (of flowmeters),
   141
static pressure, 62
stator (of synchros), 92
Stefan's constant, 53
step index fibre, 241
step response,
   first order system, 15
   second order system, 18
stiction, 11
still well, 193
Stolz equation, 130
straightening vanes, 132
strain, 161
strain gauge accelerometer, 273
strain gauges,
   foil gauges, 167
   gauge factor, 166
   mounting, 169
   principle of, 165
   semiconductor, 170
   stress and strain, 161
   temperature compensation, 173
strain weigher, 160
streamline flow, 127
stress, 161
synchros,
   control transformer, 98
   control units, 91
   differential units, 101
   identification, 104

   general, 90
   lead transposition, 95
   receiver, 94
   synchro-to-digital converter, 107
   torque units, 91
   transmitter, 92
   zeroing, 95

torr, 82
tachogenerator, 257
taring, 187
temperature compensation of strain
   gauges, 173
temperature measurement,
   bimetallic thermometers, 23
   expansion thermometers, 21
   principles of, 21
temperature scales, 20
thermistors, 32
thermocouples,
   extension and compensating cable,
     42
   general, 36
   tables, 41
   thermocouple laws, 37
   types, 41
thermopile, 55
thermostats, 23
thermowell, 42
Thévenin equivalent of Wheatstone
   bridge, 178
Thompson weir, 158
three- and four-wire connection,
   resistance thermometers,
     32
throat tap, 133
time constant, 16
time delay (pneumatic), 79
torque measurement, 179
torque transmitter, 91
transducer, 3
transmissive mode (of LCD), 236
transmitter, 3
transmitter (synchro), 92
triple point of water, 20
turbine flowmeter, 142
turbulent flow, 127
turndown ratio, 142
two-wire 4–20 mA transducer, 4

V compensating cable, 40
V-notch weir, 158
vacuum gauges, 82
vapour pressure thermometer, 25

variable area flowmeter, 144
variable capacitance transducers, 119
variable inductance transducer, 71, 119
variable reluctance tachometer, 264
variable reluctance transducer, 71
velocity,
    drag cup, 260
    pulse tacho, 261
    relationships, 256
    tachogenerator, 257
velocity (of flow), 125
velocity profile, 127
vena contracta, 129
venturi tube, 133
vernier Moiré fringe encoder, 114
vibration transducers, 277
virtual image, 221
viscosity (kinematic), 128
visible light, 216
voltage-to-current conversion, 5
volumetric flow, 125

vortex shedding flowmeter, 146

wavelength,
    electromagnetic radiation, 51, 216
    sound (ultrasonic), 265
weighers,
    force balance, 159
    loadcells, 183
    magnetoelastic, 181
    strain weigher, 159
    types of, 159
weight controllers, 187
weir (flow measurement), 158
Wheatstone bridge, 170

X-rays, 216

Young's modulus, 163

Zeroing,
    of bridge circuits, 176
    of synchros, 95